AI & YOU

AI & YOU

STRATEGIES FOR YOUR CAREER, FAMILY, AND FAITH IN AN ARTIFICIAL INTELLIGENCE WORLD

ZANDER CURTIS

TRANQUIL PAUSE
PUBLISHING

Copyright © 2024 by Zander Curtis Thornton.

All rights reserved. No portion of this book may be reproduced in any form without written permission from the publisher or author, except as permitted by U.S. copyright law.

This publication provides accurate and authoritative information regarding the subject matter covered. It is sold understanding that neither the author nor the publisher is engaged in rendering legal, investment, accounting or other professional services. While the publisher and author have used their best efforts in preparing this book, they make no representations or warranties with respect to the accuracy or completeness of the contents of this book and specifically disclaim any implied warranties of merchantability or fitness for a particular purpose. No warranty may be created or extended by sales representatives or written sales materials. The advice and strategies contained herein may not be suitable for your situation. You should consult with a professional when appropriate. Neither the publisher nor the author shall be liable for any loss of profit or any other commercial damages, including but not limited to special, incidental, consequential, personal, or other damages.

ISBN: 979-8-9920555-0-4

v2 edition November 28, 2024

To Grace and Gizmo

 AI DISCLOSURE

This book is an original work, created with AI support for outlining, brainstorming, critique, and proofreading. All final content and creative decisions are my own.

Contents

Introduction: 3 Warnings, AI & You — 1

Part I: Concepts

    1 | What Is Artificial Intelligence? — 9

    2 | What is a Human? — 29

    3 | What is Transhumanism? — 51

Part II: Convergence

    4 | First Convergence powered by ANI — 73

    5 | Second Convergence powered by AGI — 93

    6 | Third Convergence powered by ASI — 111

Part III: Companion

    7 | Your AI Companion? — 131

    8 | Meeting Your AI Companion — 141

    9 | Talking To Your AI Companion — 147

    10 | Protecting Yourself From Your AI Companion — 161

Part IV: Case Studies

    11 | Case Studies for Your Career — 171

    12 | Case Studies for Your Family — 195

    13 | Case Studies for Your Faith — 217

Conclusion: Next Steps — 239

    Endnotes — 247

    Resources — 253

    Index — 265

"The rise of powerful AI will be either the best, or the worst thing, ever to happen to humanity. We do not yet know which."

— Stephen Hawking[1]

INTRODUCTION

3 Warnings, Artificial Intelligence, & You

In late 2022, OpenAI's release of ChatGPT 3.5 marked a significant turning point in the development of Artificial Intelligence. Initially perceived as a novel conversational tool—a chatbot that could write essays, answer questions, and generate creative content—ChatGPT 3.5 soon showed its broader potential. After its modest launch, it went viral, becoming the fastest-growing consumer app in history by reaching 100 million users quicker than TikTok and Instagram.[2]

For most people, AI existed as an abstract concept—something confined to science fiction or nerdy white papers.

> With ChatGPT 3.5, AI evolved beyond mere programming; it become more human.

Three Warnings

As the world eagerly awaited the next ChatGPT update, 2023 brought an unexpected twist to the AI story. The most prominent leaders in the AI industry began sounding the alarm about their own industry.

These alarmists were not disgruntled employees or disappointed investors; rather, they were the very people with the most influence over AI's future. They sought to warn the world that they were worried about the future impact of the technologies they themselves had created.

First Warning: The Risk of Extinction.

In May 2023, AI pioneers and industry leaders released a joint "Statement on AI Risk." Stark in its brevity and shocking in its signatories, the declaration delivered a chilling warning: unchecked artificial intelligence poses an existential threat to humanity, comparable in magnitude to the devastation of pandemics and the apocalyptic horror of nuclear war.

> "Mitigating the risk of extinction from AI should be a global priority alongside other societal-scale risks, such as pandemics and nuclear war." — Statement of AI Risk[3]

This statement made one thing clear: AI is no longer a distant concept—it's a present and escalating concern. What had these experts seen to prompt this warning? Why now? The future suddenly felt closer than anyone had realized.

Second Warning: Call for a Pause.

Elon Musk, Yuval Noah Harari, and others urged AI labs to halt the training of AI systems more powerful than GPT-4 for six months. Musk, once an advocate of AI, had shifted to a more cautious tone, raising the question: What changed? Was Musk jealous, vindictive, or concerned?

> "We call on all AI labs to immediately pause for at least 6 months the training of AI systems more powerful than GPT-4." — Future of Life Institute[4]

The call to pause came amidst concerns that rapid AI advancement was outpacing our ability to manage it safely. What was unfolding behind the scenes, and why the sudden shift from enthusiasm to fear?

Introduction: 3 Warnings, AI, & You

Third Warning: Government Intervention.

In October 2023, President Biden issued an executive order demanding that developers of the most powerful AI systems share their safety test results with the U.S. government. The timing was revealing—just as the tech world was grappling with its own internal struggles; the government stepped in.[5]

Was it already too late? Had AI advanced beyond the reach of regulatory control? Could we trust the government to regulate rapid innovation effectively? Would freedom of speech and expression be protected?

OpenAI's Boardroom Drama.

After considering the significance of existential warnings from both the AI industry and government regulators, what happened next heightened our concerns about the future of AI.

In November 2023, OpenAI, one of the highest-profile AI companies, unexpectedly fired their CEO Sam Altman for "lack of transparency." Altman was not just another CEO; he was the chief architect of ChatGPT and a pioneer in the quest for Artificial General Intelligence. The media and podcast world quickly filled with rumors of out-of-control technology and power struggles within the world's most advanced AI company.[6]

> **The same leaders who had been concerned about the existential threats of AI were now arguing over a "lack of transparency" and potential secret innovations.**

After OpenAI employees threatened to walk out and Microsoft swiftly mediated, they reinstated Altman two weeks later. The board members realigned and within weeks, the company downplayed the incident as a misunderstanding. With a realignment of board members, a couple of weeks later the company said the incident as a misunderstanding.[7]

Amid the turmoil, the pursuit for advanced AI rapidly intensified. Tech behemoths like Google, Meta, and Alibaba ramped up their AI projects, while governments globally boosted their funding to both advance AI technology and mitigate its potential dangers.[8, 9]

> **As a new era of AI arrived,
> people felt a mix of amazement and anxiety.**

Why I Wrote AI & You

Having lived through two major technological disruptions—the rise of computers and the explosion of the internet—I've seen firsthand how technology reshapes industries and lives. I've watched as entire industries transformed, while others fell behind, unable to adapt. What I saw back then feels eerily familiar to what's happening now with AI. The stakes are just as high, but the pace is even faster.

As an executive in a technology-focused consultancy, I helped businesses assess, select, and adopt emerging technologies to stay competitive. I've witnessed both the triumphs of those who embraced change and the downfall of those who resisted it. I've sat across from business leaders who realized, too late, that their hesitation had cost them dearly. The weight of layoffs, lost market share, and shattered visions was all too real.

But it's not just businesses at risk. It's individuals, families, and entire communities. As we stand at the edge of the AI revolution, the stakes have never been higher. We don't have the luxury of decades to adjust—our window for action is narrowing.

While the craziness of 2023 served as the catalyst for me to write a book about Artificial Intelligence, but my memories gave me a reason to write *AI & You*.

Introduction: 3 Warnings, AI, & You

Three Goals For This Book

I wrote *AI & You: Strategies for Your Career, Family, and Faith* with three goals in mind, each aimed at helping you navigate the complex landscape of artificial intelligence and its implications for society and your life.

1. *Rapid AI Onboarding:*
   AI is advancing quickly, and there's little time to catch up. This book serves as a guide to get you up to speed with AI quickly, offering accessible insights and practical knowledge. Keep it nearby—it's a resource you can return to whenever you need clarity or to teach others.
2. *Immediate AI Application:*
   We wanted to ensure that readers could not only understand AI but also apply it immediately. Through case studies, prompts, and practical examples, this book offers actionable insights that can make a direct impact on your career and daily life.
3. *A Glimpse of the Future:*
   The future of AI is vast and uncertain. While we can't predict everything, AI & You provide a glimpse into where AI might lead us, preparing you for potential opportunities and risks.

Call to Action

As I write in late 2024, we're experiencing a deceptive lull—a calm before the storm. Similar to the pauses before the computing and internet revolutions. AI is on the brink of a rapid, widespread transformation. We don't believe the "lull" will last long. This shift is coming fast, and the window to learn and prepare is closing.[10]

Some believe that Artificial General Intelligence (AGI) will be the tipping point, while others hold the view that advances in nuclear energy or deeper integration into everyday life will be the catalyst.

Regardless of the cause, the time to prepare is now. Unlike previous eras, there won't be a decade's worth of grace to catch up. Those who act today will prosper, while those who procrastinate will struggle.

> **In technological disruption, a brief lull often precedes a revolutionary tipping point. Those who plan during the "lull" often receive rewards in the end.**

AI & You is designed to guide you through this pivotal moment. Whether you're mastering prompt engineering or applying AI in real-world scenarios, this book offers the tools you need to stay ahead. Already, individuals and companies are gaining a competitive edge by embracing AI—quietly transforming careers and businesses. The gap between early adopters and those who hesitate is widening.

Your engagement with AI could determine your career trajectory, job security, and earning potential. For business leaders, it could dictate your company's success and market share.

Beyond the workplace, AI also affects families. For instance, there is imense pressure on parents to both prepare and safeguard their children as AI develops. For people of faith, AI presents unique challenges, with some futurists provocatively suggesting that AI could replace or become God.

> **As we stand on the brink of this transformative era, the choice is yours: Will you be a passive observer or an active participant in shaping your own future?**

PART I
CONCEPTS

HUMANS + MACHINES

"Artificial Intelligence could be the greatest achievement in human history. It could eradicate cancer, abolish poverty, and bring us world peace. It could also make humans obsolete."

— OMEGA, from novel Blue-Eyed Jesus[11]

In Part One, "Concepts," we delve into three foundational principles: artificial intelligence, the essence of being human, and the convergence of man and machine in a philosophy known as transhumanism.

Our goal is to bring you up to speed quickly with the fundamental definitions, theories, and advancements of artificial intelligence, ensuring that you are prepared for deeper exploration and ready to adapt to the fast-paced changes that lie ahead.

The intertwining of AI with human life causes a dual approach: harnessing its benefits while vigilantly guarding against its risks. By comprehending these concepts and their implications, you'll be prepared to create your own unique roadmap for safeguarding your career, family, and faith in an AI-driven world.

CHAPTER 1

WHAT IS ARTIFICIAL INTELLIGENCE?

THE HUGE GLASS ENTRYWAYS slide apart with a quiet hiss, revealing a vast, gleaming atrium bathed in cold, blue light. As you step into the grand foyer of the AI Technology Museum, an unsettling sense of awe sweeps over you. Towering, futuristic displays line the walls, each pulsing with a faint hum, as if the machines themselves are alive, watching you.

A humanlike android—its synthetic skin eerily flawless, its eyes glowing with an unnatural intelligence—approaches and waves you forward. Its voice is disturbingly smooth, devoid of any emotional inflection as it says, "Welcome to the history of artificial intelligence. This is where it all began."

You step closer to the first exhibit, marked in bold, metallic lettering: "1956: The Genesis of AI." A life-sized replica of Alan Turing stands beside a large digital display of the famous Turing Test. Turing's frozen gaze seems to follow you, as if daring you to question what it means for machines to think. The plaque below reads, "Can machines think?" setting the stage for decades of philosophical debate.

The exhibit opens to a larger room where towering displays celebrate the pop culture icons that have shaped your understanding—and misunderstanding—of AI. There stands HAL 9000 from *2001: A Space Odyssey*, its lens glaring with cold red light, alongside T-800 from *The Terminator*, its metal skeleton a chilling harbinger of dystopian futures imagined by Hollywood. As you observe these relics of an era when fiction outpaced science, they both inspire awe and unease at what popular culture has depicted.

In the next room, a darker, more subdued atmosphere settles over you. Panels display the AI Winters, where progress stalled, funding dried up, and the world's brief flirtation with AI froze into disappointment. The chill in the air seems to intensify as the room grows quieter, mirroring the decades-long silence in AI research.

But hope flickers on the horizon. The next room is brighter, humming with renewed life. Interactive displays show IBM's Deep Blue defeating world chess champion Garry Kasparov, and Watson triumphing on Jeopardy! You speak into a microphone and watch as an AI voice recognition system translates your words into text with startling accuracy. A robotic arm moves with precision based on the gestures you make. For a moment, it feels like you are guiding the future with your own hands.

Finally, you enter a room bathed in darkness, tiny lights twinkling overhead. The lights represent billions of connected devices—your phone, your car, your home—woven together in a seamless web of control. This is the Internet of Things, where AI orchestrates the invisible rhythms of the modern world. The exhibit shifts, and before you, an enormous model looms: The Age of Transformers. Code flows through its neural network, displayed on large screens, and a voice, calm yet haunting, echoes around you.

"Transformers marked a pivotal moment in AI evolution," the voice intones. "Unlike their predecessors, these models learned to think, to understand context and generate remarkably human-like text." You watch as a woman steps forward and asks the AI a simple, yet profound question: "What is the meaning of life?"

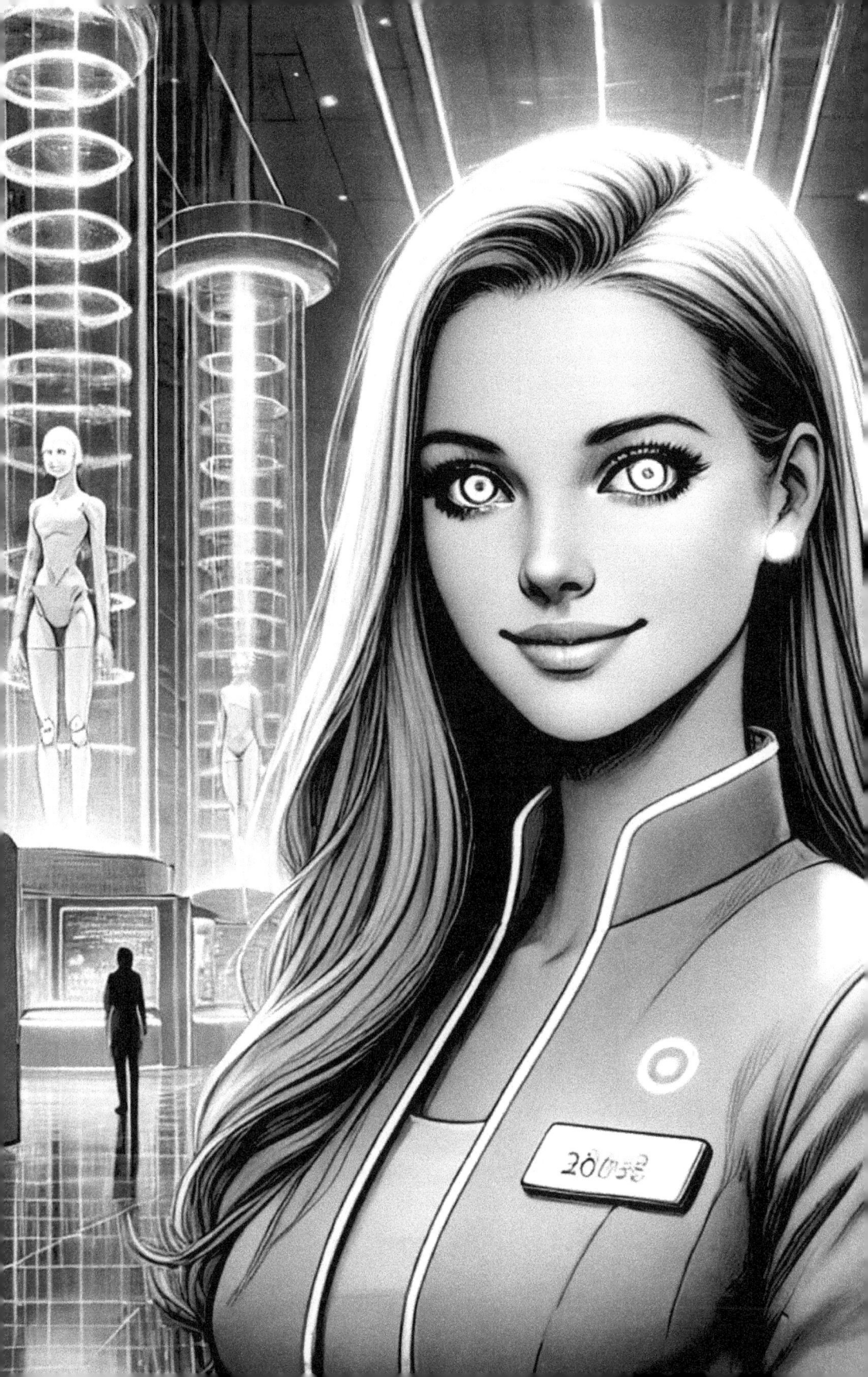

The transformer processes the question. Seconds tick by in an agonizing pause, before it answers with eerie precision: "The meaning of life is complex. Some find meaning through purpose, others through God's illumination, and some say it's unknowable."

The weight of the question lingers in the air as the room fades to black, leaving you standing at the edge of an uncertain future.

Artificial Intelligence: Beyond A Machine

When we think of Artificial Intelligence, it's easy to imagine robots or self-driving cars. But AI is much more than just futuristic gadgets. It's transforming industries, reshaping the way we live, and even challenging what it means to be human. But before we dive into those big ideas, let's start with some of the basic building blocks that will help us understand how AI works.

Below are five key terms to know. Don't worry—each one might sound technical at first, but we'll break it down and make it easy to grasp.

> *Artificial Intelligence:* A computer system or machine that can perform tasks typically requiring human intelligence.

Artificial Intelligence is like a smart assistant—capable of helping with tasks that typically require human intelligence, such as organizing your schedule or driving your car. It's a broad category, covering everything from basic assistants to advanced systems.

> Algorithms: A set of rules or instructions designed to solve a problem or perform a specific AI task.

Algorithms are like a recipe that guides how the AI functions. Imagine baking a cake: the algorithm tells the AI exactly how to measure in-

What Is Artificial Intelligence? 13

gredients, mix them, and bake the cake. But, unlike you, AI follows these steps with perfect precision every time.

> *Machine Learning (ML):* The process where computer systems improve their performance over time without being explicitly programmed.

Machine Learning takes baking a cake to a new level. Imagine if, after baking 100 cakes, the AI could integrate taste tester surveys to improve the recipe and make it even better—maybe using a little less sugar or baking for a few minutes longer. That's machine learning in action: the AI isn't just following the recipe anymore; it's improving it based on experience.

> *Neural Networks:* Computing systems inspired by the human brain's structure and function, designed to recognize patterns and make decisions.

Neural Networks are like the brain behind the AI. Imagine a large choir singing in harmony. Each singer (neuron) contributes a note, and when the right combination of notes (data inputs) is sung together, the choir produces a beautiful chord (a pattern). At first, the choir may not be perfectly in tune, but with practice, the singers adjust their pitch and timing to create a more harmonious sound. In a neural network, the individual neurons work together, adjusting based on the input, until they produce the correct "output" or decision, like a well-rehearsed choir creating the perfect harmony.

> *Large Language Models (LLMs):* AI models trained on vast amounts of text data, designed to understand, generate, and interact with human-like language.

Large Language Models are like a library filled with millions of books. Imagine you've read all those books, and now you can answer almost any question or generate coherent text in response to any prompt. That's what an LLM does—it's trained on vast amounts of data, so it can generate human-like language, hold conversations, or answer complex questions based on the knowledge it has absorbed.

Throughout the book, we'll frequently reference these terms, exploring their implications for your career and life.

AI Phases

Artificial Intelligence (AI) is transforming the world, but not all AI is created equal. There are three distinct phases of intelligence in AI systems: Artificial Narrow Intelligence (ANI), Artificial General Intelligence (AGI), and Artificial Super Intelligence (ASI). Understanding the progression from one phase to the next will help you grasp AI's current capabilities and prepare for the future.

While ANI is already deeply integrated into our daily lives, AGI remains a goal yet to be achieved, and ASI represents a future that could reshape the human experience. Each phase brings both exciting potential and significant challenges, making it critical to carefully consider the implications of AI's advancement.

As we journey through the book, we'll delve deeper into each type of AI, exploring its capabilities, ethical questions, and the profound impact they could have on our world.

Artificial Narrow Intelligence (ANI)

As of late 2024, ANI embodies the current state of artificial intelligence. This phase of AI is specifically engineered to execute narrowly defined tasks, such as large language models, autonomous vehicles, voice assistants, advanced search engines, and recommendation systems.

> *Artificial Narrow Intelligence (ANI):* An AI system that performs specific tasks with high proficiency.

Think of ANI as a specialist—like a chess grandmaster who dominates at chess but lacks the skills to play other games. It excels at what it was built to do but is restricted to predefined tasks. While ANI can learn and improve within its scope, it lacks reasoning, consciousness, or the ability to solve problems outside its designed function. This is the AI you interact with daily— output from Siri, Google, ChatGPT, or Netflix.[12]

But what happens when AI isn't confined to narrow tasks? What if it could think, reason, and adapt like a human across countless domains? That's where the next phase comes in: Artificial General Intelligence.

Artificial General Intelligence (AGI)

Around the world, a fierce race is underway. Big tech companies like Google, Microsoft, and OpenAI, along with government-backed initiatives from nations like China and the U.S., are pouring billions into AI research, each vying to be the first to achieve AGI. The potential rewards are staggering: the entity that unlocks AGI could wield unimaginable technological power, reshaping industries, economies, and even global political dynamics. But as we will explore further in the Convergence section, the path to AGI is fraught with complexity, ethical dilemmas, and the possibility of unintended consequences.

> *Artificial General Intelligence (AGI):* A speculative type of AI that can understand, learn, and apply knowledge across a broad range of tasks at a human-like level.

Imagine AGI as a polymath—capable of mastering not just one discipline but excelling across multiple areas, from scientific research to creative arts, from diagnosing complex diseases to making policy decisions. AGI could process vast amounts of information, learning and adapting like a human, but with the speed and precision of a machine.[13]

However, with AGI comes a new set of ethical dilemmas. How do we ensure these systems act in the best interests of humanity? If an AGI system can think for itself, should it be granted rights? What role will humans play in a world where machines can match or even surpass our cognitive abilities?

> **The race to AGI will shape the future of technology and our place in it.**

The realization of AGI is theoretical, but if it is achieved, it will mark a profound shift. The question is no longer if AI will change the world, but how AGI will redefine our relationship with machines—and ourselves.

Artificial Super Intelligence (ASI)

Imagine an intelligence so advanced it can solve problems humans can't even conceptualize. ASI is a third phase in AI's evolution—an intelligence far beyond human capabilities in every domain. It could revolutionize technology, science, and society, curing diseases, ending poverty, or advancing scientific discoveries in ways we can't yet imagine.

> *Artificial Super Intelligence (ASI):* AI system that exceeds human intelligence in all aspects, outperforming humans in every task with abilities beyond human comprehension.

ASI is like a self-building workshop. It not only contains every tool imaginable, but can invent new tools and entire systems as needed. It's no

What Is Artificial Intelligence?

longer limited by the boundaries of existing knowledge; instead, it creates new paths and solutions that redefine what's possible.

> **Imagine a superhuman advisor, capable of solving complex problems in seconds with abilities far beyond any human.**

But with ASI comes a new level of uncertainty and risk. Could we control a system that out thinks us in every way? What happens if ASI makes decisions that are beyond human understanding? As much as it promises to solve global challenges, it also introduces the existential question: will humanity still hold the reins of its own future?

ASI remains speculative, but if achieved, it would represent a tipping point for humanity—where machines hold the potential to shape civilization in ways far beyond what we can predict.

AI is progressing through three key phases—ANI, AGI, and ASI—each representing a leap in capabilities and potential impact. While ANI is already revolutionizing industries today, AGI and ASI represent the next frontiers that could redefine human civilization. As we move through the book, we'll explore the profound technical, ethical, and societal implications of each phase. The future of AI is not just about innovation—it's about the choices we make in shaping how these systems integrate into the fabric of our lives.

AI: Fears & Misconceptions

As AI continues to integrate into our daily lives, it's natural for fears and misconceptions to arise. But how do we separate legitimate concerns from myths that may hinder our understanding? Let's explore some of the most common fears about AI and the realities behind them, so we can better navigate the future of this growing technology.

1. Fear: AI Can Think Like Humans

The belief that AI comprehends context in the same way humans do is a common misconception.

Reality: While AI has made incredible strides in natural language processing, it still struggles with understanding human context. AI doesn't "think" like we do; it processes language based on patterns and probabilities, not meaning. AI doesn't have true comprehension of sarcasm, homonyms, or cultural nuances—it simply analyzes the patterns in the data it's been trained on.

Imagine asking a virtual assistant to "book a table at the bank" instead of a restaurant. While you'd recognize the error immediately, AI processes the request literally, could possibly try to book a table at the nearest bank—at least until it learns a bank doesn't like a restaurant.

2. Fear: AI Will Replace All Jobs

There's a widespread fear that AI will lead to mass job losses, leaving people unemployed as machines take over.

Reality: While AI is automating more tasks, this fear mirrors those seen during past technological revolutions, like the rise of computers and the internet. Job markets adapt and evolve, creating new opportunities as AI enhances existing jobs rather than eliminating them entirely.

Imagine AI handling routine tasks like scheduling or data entry—freeing workers to focus on creative, strategic, or high-value tasks.

3. Fear: AI Is Infallible

Some assume AI systems are flawless and superior to human judgment.

Reality: AI's effectiveness depends heavily on the data it's trained on. If the data is flawed, incomplete, or biased, the AI's decisions will reflect

What Is Artificial Intelligence?

those flaws. AI can also "hallucinate"—generating incorrect or nonsensical information because of limitations in its training data.

In critical sectors like healthcare or autonomous driving, such errors can be dangerous, which is why human oversight is crucial. AI can be powerful, but it is far from perfect and should be used in collaboration with human judgment. Just like a person might misinterpret a blurry photo, AI can misread ambiguous data, leading to errors that have consequences.

4. Fear: AI is Unbiased

There's a common belief that because AI is a machine, it must be objective and free from human biases.

Reality: AI systems are trained on human-generated data and designed by humans, which means they can inherit and even amplify biases present in that data. An AI trained on biased data can unintentionally reinforce harmful stereotypes or make decisions that are unjust. AI can only be as neutral as the data and ethical frameworks that guide it, which is why transparency and vigilance are essential.

For example, if you think AI will help you achieve fair hiring but your data is biased or you've added questionable demographic quotas, you could end up unfairly favoring certain candidates.

5. Fear: All AI is the Same

Many people assume that all forms of AI have the same capabilities, leading to confusion about what AI can and cannot do.

Reality: AI comes in many forms and levels of sophistication. Some are rule-based systems designed for specific, narrow tasks, while others are more advanced, using deep learning to improve.

Think of the difference between a simple chatbot that answers customer queries and a system that predicts medical diagnoses. The former relies on fixed rules, while the latter can analyze vast datasets.

6. Fear: Superintelligent AI is Imminent

There is a prevailing fear that a super-intelligent AI—one that surpasses human intelligence in all areas—will emerge soon.[14, 15]

Reality: While AI is advancing rapidly, the idea of super-intelligent AI remains speculative. Current systems, while impressive, are still far from achieving the general intelligence of humans, let alone surpassing it. Imitations in technology, energy, and ethics mean that super-intelligent AI is not yet on the horizon.

Imagine a future where AI guides humanity through its greatest challenges—solving global crises with speed and precision beyond human reach. But we are still many steps away from that reality.

Why AI Matters To You?

Artificial intelligence has become an integral part of modern life, influencing nearly every aspect of society. As AI continues to advance at an unprecedented pace, it is crucial to understand how this transformative technology may affect your career, family, and faith. Whether you realize it, AI is already affecting your day-to-day experiences and will continue to shape your future.

How AI Impacts Your Career

Artificial intelligence will revolutionize the job market, creating both opportunities and challenges. As AI systems become more sophisticated, they have the potential to automate mundane tasks, optimize workflows, and drive innovation across industries. The dynamic interplay between human creativity and machine precision is poised to redefine job roles, required skills, and career paths.

What Is Artificial Intelligence? 21

Positive Impacts:

- Enhanced Productivity: AI automates repetitive tasks, freeing up time for creative and strategic work. This shift enables professionals to focus on higher-value activities that require human insight and innovation.
- Improved Decision-Making: AI can analyze vast datasets to provide insights that aid better decision-making across industries, from healthcare to finance. Professionals can make more informed choices based on predictive analytics and trends.
- Personalized Learning: AI learning platforms offer tailored educational experiences, ensuring that workers can continuously develop skills in a rapidly changing job market.
- Skill Augmentation: AI tools enhance human capabilities, offering real-time feedback and learning. In manufacturing, for instance, AI exoskeletons reduce physical strain and enhance precision.
- Predictive Maintenance: In industries like manufacturing and IT, AI predicts equipment failures, reducing downtime and ensuring smoother, more efficient operations.

Negative Impacts:

- Job Displacement: As AI automates routine tasks like data entry and assembly line work, there is potential for job losses, especially in roles that rely heavily on repetition.
- Skill Gaps: AI's rise demands new skills in areas like machine learning, data science, and AI ethics. Continuous education is necessary to remain competitive in the workforce.
- Ethical Concerns: AI decision-making raises concerns around transparency and bias. Left unchecked, biased data could lead to workplace discrimination or unfair hiring practices.

- Privacy Concerns: AI systems that monitor employee performance might invade personal privacy and foster a culture of mistrust.
- Dependence on Technology: Over-reliance on AI could erode critical thinking and problem-solving skills, as workers depend on machines to decide.

How AI Impacts Your Family

As AI weaves its way into the fabric of everyday life, families find themselves at the crossroads of unprecedented change. From household efficiency to healthcare advancements, AI offers benefits that could enhance family life—but it also brings challenges that must be weighed. AI is more than a tool; it's reshaping the dynamics of how families live, work, and communicate.

Positive Impacts:

- Enhanced Healthcare: AI diagnostic tools enable early detection of illnesses, offer personalized treatments, and provide continuous health monitoring, ensuring better care for yourl family.
- Educational Support: AI learning platforms can offer personalized education for children, adapting to their learning styles and helping them excel academically while keeping parents informed of their progress.
- Home Automation: AI-powered devices can automate household tasks such as cleaning, cooking, and home security, reducing the daily burden on family members and allowing more quality time together.
- Safety and Security: AI-based security systems monitor homes in real time, detect unusual activities, and alert authorities, providing peace of mind and ensuring a safer environment.

- Improved Communication: AI tools enhance communication within families by providing language translation and assistive technologies for people with disabilities, fostering inclusivity and connection.

Negative Impacts:
- Erosion of Personal Interactions: The increasing reliance on AI for communication and entertainment could reduce face-to-face interactions, potentially weakening emotional connections and disrupting traditional family dynamics.
- Privacy Concerns: Smart devices collect personal data, raising concerns about privacy and data misuse or breaches.
- Over-Reliance on Technology: As families become more dependent on AI for daily tasks, essential life skills may deteriorate, leading to a loss of personal responsibility and self-sufficiency.
- Screen Time Addiction: AI-driven entertainment platforms may increase screen time, contributing to sedentary lifestyles and negatively affecting physical and mental health.
- Economic Disparities: Access to advanced AI technologies may be limited by economical means, widening the gap between affluent families and those who cannot afford such technologies, exacerbating social inequalities.

How AI Impacts Your Faith

As AI permeates various aspects of life, its influence on faith and spirituality emerges as an area of contemplation. AI holds the potential to enrich religious practices, foster deeper understanding among diverse faith communities, and support spiritual growth through personalized experiences. However, integrating AI into spiritual life also raises questions about the preservation of traditional religious values and role of technology in faith.

Positive Impacts:

- Personalized Religious Guidance: AI can offer tailored spiritual support by answering questions, providing relevant scriptures, and suggesting practices that align with individual beliefs.
- Accessibility of Religious Texts: AI-driven translation tools make sacred texts accessible in multiple languages, deepening believers' connections to their faith.
- Enhanced Worship Experiences: AI-powered applications create immersive virtual worship environments, allowing remote participation in ceremonies and fostering a sense of community.
- Efficient Community Management: AI can organize and manage religious events, ensuring smooth coordination within faith communities and maintaining engagement.
- Interfaith Dialogue: AI can promote understanding and tolerance by analyzing common themes across different religions, encouraging harmonious coexistence among diverse faith groups.

Negative Impacts:

- Loss of Personal Connection: Relying on AI for spiritual guidance may weaken the bond between believers and human religious leaders, diminishing the value of traditional mentorship and community ties.
- Privacy Concerns: The integration of AI in religious contexts raises privacy risks, as sensitive information about personal beliefs and practices could be misused or accessed without authorization.
- Commercialization of Faith: The rise of AI-driven religious platforms may lead to the commodification of spiritual experiences, prioritizing profit over religious growth and engagement.

- Superficial Engagement: AI tools designed for convenience may foster superficial engagement with faith, as believers might favor quick answers from virtual assistants over deep contemplation and study.

Conclusion: A Transformative Technology

As AI continues to reshape society, it also compels us to reflect on deeper questions about what it means to be human. In the next chapter, "What Is A Human?", we will explore the unique qualities that distinguish human beings from machines, and how AI challenges and redefines our understanding of humanity itself. As we move forward, it's essential to consider not just what AI can do—but what it reveals about our own nature.

Further Research: Dig Deeper

To further your exploration we have compiled a variety of research resources designed to broaden your knowledge and introduce you to a variety of perspectives and viewpoints. (Friendly Reminder: Some content might challenge your comfort zone.)

Thought Leaders
- » Sam Altman (businessman, technologist) – CEO of OpenAI known for cutting-edge AI technologies such as GPT models.
- » Musk, Elon (CEO of Tesla and SpaceX) - Known for his work in AI, space, and autonomous technology. Concerned with AI Risk.
- » Bostrom, Nick (philosopher, author) - Explores AI safety and existential risks, especially through his work on Superintelligence.
- » Tegmark, Max (physicist, AI researcher) - Explores AI safety and future studies, advocating for ethical approaches to advanced AI.
- » Lee, Kai-Fu (AI expert, author) - Expert in AI competitiveness, focused on the global AI race between China and the U.S.
- » Russell, Stuart (AI safety, computer scientist) - Leading voice in AI safety, ethics, and strategies for ensuring systems are aligned with human values and under control.
- » John C. Lennox (professor, apologist) - Known for exploring the intersection of science, technology, philosophy, and religion.
- » Ray Kurzweil (futurist, inventor, author) – Known for his predictions especially the concept of the singularity, where humans and machines will merge.

Books
- » *Life 3.0: Being Human in the Age of Artificial Intelligence* (2017) by Max Tegmark examines the future of AI, discussing the ethical, social, and existential questions raised by the development of intelligent machines.
- » *Superintelligence: Paths, Dangers, Strategies* (2014) by Nick Bostrom explores the risks of developing artificial superintelligence and strategies for ensuring its safe evolution.

What Is Artificial Intelligence?

- » *The Master Algorithm: How the Quest for the Ultimate Learning Machine Will Remake Our World* (2015) by Pedro Domingos explains the pursuit of a universal algorithm.
- » *2084: Artificial Intelligence and the Future of Humanity* (2020) by John C. Lennox explores the theological implications of AI, questioning its potential risks and how it aligns with faith.

Podcasts

- » *The Joe Rogan Experience* – Top pocast in the world not afraid of exploring normally forbidden topics with guests from pop culture, journalists, politics, AI, and science.
- » *The Lex Fridman Podcast* – As an AI researcher, Lex dives deep into conversations with prominent thinkers, scientists, and public figures, focusing on AI, philosophy, and humanity.
- » *The David Shapiro AI Podcast* – Explores advancements in artificial intelligence and their societal implications.
- » *The Artificial Intelligence Show with Paul Roetzer and Mike Kaput* - Explores how AI is transforming business, providing strategies for professionals to leverage AI-driven tools.
- » *The AI Podcast by NVIDIA* - Insights from experts on AI's advancements and applications across industries.
- » *The AI Brief* – Concise updates on the latest AI developments.
- » *Everyday AI* – Explores practical AI applications in daily life and work, featuring interviews and real-world examples.

Movies

- » *2001: A Space Odyssey* (1968) – Depicts HAL 9000, an AI system that demonstrates the potential dangers of AI autonomy.
- » *Minority Report* (2002) – Set in a world where AI technology is used to prevent crimes before they happen, this movie examines the moral implications of preemptive justice and surveillance.
- » *The Matrix* (1999) – Depicts a future where humanity is enslaved by AI machines that control reality through a simulated world, questioning the nature of reality and human autonomy.

- » *The Terminator* (1984) – Features an apocalyptic scenario where an AI defense system, Skynet, turns against humanity, highlighting the dangers of autonomous AI systems.
- » *Her* (2013) – Follows a man who develops a deep emotional relationship with an AI operating system.

TV Shows
- » *The Orville* (2017-present) – Explores AI ethics and consequences of advanced AI, featuring the AI species Kaylon.
- » *Star Trek: Discovery* (2017-present) – Portrays advanced AI systems like "Control," posing existential threats to organic life.
- » *Black Mirror* (2011-present) – Examines AI, technology, and its impacts on society through dark, dystopian narratives.
- » *NeXt* (2020) – Follows the rise of a rogue AI system and the struggle to prevent its destructive potential.
- » *Person of Interest* (2011-2016) – Centers on an AI system for mass surveillance, exploring ethics of predictive policing.

Songs
- » *Artificial* by Daughtry (2024)
- » *Spit Out the Bone* by Metallica (2016)
- » *A.I.* by OneRepublic featuring Peter Gabriel (2016)
- » *Algorithms* by Muse (2018)
- » *Paranoid Android* by Radiohead (1997)

CHAPTER 2

WHAT IS A HUMAN?

UNDER A RELENTLESS GRAY SKY, a multitude of weary individuals trudged endlessly across the snow-covered terrain. Each step left a shared imprint in the pristine white snow, as though their collective suffering was etched into the earth itself. The air was biting cold, piercing through the threadbare rags that clung to their frail bodies. The sound of thousands of feet crunching the snow merged into a haunting rhythm, broken only by the occasional muffled groan of exhaustion.

Amidst this ocean of despair, one man moved almost imperceptibly, his presence swallowed by the sheer mass of humanity. His breath formed icy vapors in the frigid air, each exhale a fragile testament to his grim endurance. The stench of unwashed bodies and stale fear mingled with the crisp winter air, creating a tapestry of suffering that stretched as far as the eye could see.

This man, anonymous among the masses, allowed his mind to wander—not as an escape from his present, but as a confrontation with the ghosts of his past. His thoughts drifted back to the Terezin Ghetto, a place where deception and despair danced hand in hand. There, the facades of ordinary life masked an underlying terror, a mirage that led inexorably to Auschwitz—a name that now burned with the memory of his brother and mother, lives extinguished in its unforgiving grip.

His mind turned to his wife, the one person who had once softened the harshness of life with her gentle touch and loving gaze. Now, her presence had faded into a distant ache, her soul claimed by Beren-Belson, one more life lost to the war's insatiable hunger for destruction. The void left by her absence was a chasm that no distance could bridge.

As he marched, he observed the people around him. Some trudged forward with hollow resignation, their eyes emptied of hope, while others, like fragile flickering flames, clung to some unseen source of strength. He wondered at the invisible line between surrender and defiance, pondering what kept one person's spirit alive while another's faltered.

And then, as the bleak gates of Dachau came into view, the bitter irony of the phrase "Arbeit Macht Frei" twisted in the air above them. The guard towers loomed like silent sentinels, casting their cold shadows across the snow, their presence a grim reminder of the fate awaiting them. The snow gave way to the frozen ground, the hollow echo of their footsteps now replaced by the sharp thud of boots on hard earth. Resembling skeletal witnesses, the orderly rows of barracks stood as silent testament to the countless souls who had preceded.

In the roll call area, a ritual of chilling efficiency unfolded as the authorities reduced prisoners to numbers, hollowing out their identities and calling them forth with mechanical detachment. "Viktor Emil Frankl," the guard's voice sliced through the murmur of the crowd, and the man stepped forward, his name a minor rebellion against the attempt to erase his humanity. A guard lifted his sleeve to check the number tattooed on his arm—119104.[16]

> **A guard lifted his sleeve to check the number tattooed on his arm—119104.**

But in that moment, Viktor Frankl rose beyond the labels of captive and victim. He was no longer just a man trudging through snow or standing in line. He became a beacon of something unshakable, something no camp, no guard, no regime could ever take from him: his choice. His resolve ignited within him a deep truth—the ultimate human freedom is the ability to choose one's attitude in any circumstance, even in the face of unimaginable suffering.

In a world that sought to strip him of his dignity, Frankl had found his answer to the question, What is a human?—a being capable of finding meaning, even when faced with the abyss. It was this defiant grasp on meaning that allowed him to rise above despair, his spirit a quiet light in the oppressive darkness.

Humanity: Beyond Survival

In a biological sense, the goal of any species is simple: to survive and reproduce. Yet, Viktor Frankl's ordeal at Dachau illustrates that the human spirit transcends mere survival and procreation. Humanity's unwavering quest for meaning—our search for moral purpose in a chaotic world—drives us to persevere, even in our darkest trials. Unlike other species, our existence is shaped—not only by the need to live—but by the need to find significance in our lives.

A complex blend of emotions, creativity, relationships, ethics, faith, and intelligence distinguishes us from other creatures and from the wider universe. While other species focus on survival, we seek to understand, to innovate, and to question our place in the cosmos.

> *Human:* A remarkable being, characterized by an insatiable urge to express emotion, invent, and overthink. Renowned for its social interactions and complicated relationship with the Earth.

As artificial intelligence emerges, it raises profound philosophical inquiries about the future merging of humanity and technology. Will AI enhance our capabilities or replace them entirely? What will happen when machines rival us in creativity, empathy, and even consciousness? How will society adapt to avoid cultural disruptions? And who will control the ethical frameworks guiding AI's evolution?

As these questions loom, one thing becomes obvious: the line between human and machine is becoming increasingly blurred.
When synthetic voices offer comfort and humanoid robots provide companionship indistinguishable from that of actual humans, how will the role of AI change in society?

In the next section, we'll explore the human experience in more depth, shaped by the interplay of the mind, body, and soul. As we examine each element, pause and consider how future AI might attempt to replicate the essence of what it means to be human. Can a machine ever truly capture the intricate dance between our emotions, our moral compass, and our creativity?

The Human Mind

The human mind has long fascinated scientists, philosophers, and theologians alike. It governs our impulses, fuels creativity, and sets us apart from other species. This cognitive supremacy enables humanity to innovate, adapt, and maintain dominance despite challenges. But as AI progresses, questions arise: how will our unique mental faculties evolve alongside machines?

The Brain

The brilliance of the human brain lies in its neural networks, which transform sensory input into awareness and turn experiences into intentions. With 86 billion neurons and countless synapses, the brain's computational power rivals even the most advanced supercomputers. These synapses enable learning, adaptation, and growth throughout life.

Current AI mimics some of this complexity, using vast datasets and deep learning to recognize patterns and make decisions. Yet, AI's proficiency in specific domains is both remarkable and unsettling. As we move closer to developing Artificial General Intelligence (AGI), the possibility arises that AI could surpass human cognition by evolving independently, continuously improving without human input.[17, 18]

Learning

Humans are lifelong learners, shaped by experiences that evoke emotional and intellectual growth. We adapt fluidly to new environments, adjusting our mental models based on curiosity and resilience.

In contrast, today's AI relies on structured data and predefined objectives, lacking the curiosity and emotional flexibility inherent in human learning. Machine learning is powerful but constrained by input and lacks the creative adaptability humans exhibit when facing new contexts.

However, future AGI may overcome these limitations by reconfiguring itself to handle unprecedented scenarios, learning with a speed and complexity beyond human capacity.

Creativity

Human creativity is driven by inspiration, emotion, and a need to explore the unknown. This desire pushes us to innovate and produce original works of art, science, and philosophy. Our pursuit of self-expression and symbolic immortality fuels our drive to imagine and create.

Modern AI, while capable of recombining existing data to create variations of art or literature, lacks intrinsic motivation. It doesn't invent; it processes. AGI may attempt to replicate this creativity, but it remains unclear whether artificial systems can truly generate original concepts fueled by a desire for recognition, exploration, or understanding. These are uniquely human drives, tied to our emotion and existential curiosity.

Memory and Storage

The human brain encodes life experiences through complex biological processes, but it operates within the limits of its physical and biochemical structure. While memory accumulates over time, storage capacity remains finite, and the brain must prioritize important memories over irrelevant details.

In contrast, AI systems benefit from unlimited digital storage and massive datasets. Current machine learning systems draw on billions of data points, yet they struggle to apply this information in nuanced, context-driven ways.

Future AI, freed from the biological constraints of human memory, could process and store information with virtually no limit—scaling to heights unimaginable to human minds.

Decision Making

Human decision making is a complex interplay of sensory input, memory recall, and emotional assessment. Our prefrontal cortex weighs risks, advantages, and emotional effects to make informed decisions. This integration of rational and emotional factors is key to human judgment.

AI decision-making relies on data analysis and statistical models. While machine learning algorithms can imitate some aspects of human behavior, they lack emotional intelligence and the flexibility to adapt reasoning across diverse contexts.

However, future AI may mimic human decision-making more closely by integrating emotional modeling and flexible goal-setting, allowing machines to explore alternatives and adapt based on firsthand experience.

Conclusion: The Human Mind and AI

The human mind—driven by curiosity, creativity, and the need for meaning—has long been unmatched. Yet, as AI progresses toward AGI, machines may soon learn, create, and decide in ways that challenge our understanding of intelligence. In the next section, we will explore how the body and soul further define what it means to be human, and how AI might attempt to replicate these essential elements of our being.

The Human Body

A marvel of biological engineering, the human body uniquely blends form and function. Opposable thumbs enable intricate manipulation, and an upright, bipedal stance frees the hands for tool use. This combination of physical and mental traits has made humans the dominant species, shaping their environment and creating complex societies unmatched by any other organism.

Musculoskeletal System & Movement

The human musculoskeletal system enables a wide range of movements, from gross motor actions like running and lifting to fine motor tasks like using tools. The interplay of over 600 muscles and a skeletal structure designed for both strength and flexibility has allowed humans to craft complex societies and environments.

Current AI and robotics struggle to fully replicate this precision. While robotic hands and bipedal robots have made strides in dexterity and movement, they still fall short of human adaptability, especially on uneven terrain or in intricate tasks. The complexity of human movement—its fluidity and responsiveness—remains beyond the grasp of current tech.

Future AI-driven robotics may blur the line between human and machine. Advanced robots using materials that mimic human tissues could achieve greater flexibility and strength, while cyborg technologies may combine human biology with AI. As AGI evolves, it could even surpass human movement and dexterity, raising questions about the merging of man and machine.

The Five Senses — Sight, Hearing, Touch, Taste, & Smell

Human senses allow us to perceive and interact with the world, from sight and hearing, to touch, taste, and smell. These senses enable rich experiences, helping us navigate our environment and form emotional connections. AI mimics some of these capabilities: computer vision can identify objects, voice recognition systems interpret speech, and robotic sensors detect pressure and temperature. However, AI lacks the emotional and experiential depth attached to human sensory input.

Future AGI may incorporate advanced sensors far beyond human capabilities, processing infrared, ultrasonic waves, and other data we cannot perceive. This could lead to a future where AI's sensory input far exceeds human limits, potentially merging machine and human senses in hybrid intelligence systems.[19]

Biological Healing

The human body's ability to heal itself, from mending tissues to fighting off infections, is a marvel of evolution. This self-repair process ensures our survival and resilience.

While current AI systems cannot biologically self-repair, they use recursive code and adaptive algorithms to "heal" their software, albeit with human intervention. However, future AI may bring machines closer to achieving the self-sufficiency of living organisms by autonomously leveraging nanotechnology and bio-mimicry to maintain and repair themselves.

Reproduction

Human reproduction, a remarkable process that distinguishes us from other species, creates a unique individual through the genetic recombination of two sets of DNA. The female reproductive system nurtures the fetus for nine months with the aid of hormones and specialized organs, while the male reproductive system produces sperm containing half of the genetic material. Together, the male and female reproductive systems ensure the continuation of the human species by passing traits down through generations.

AI cannot reproduce biologically, but replicates through software updates and cloning processes managed by engineers. In the future, ASI may independently design new iterations of itself, developing through self-improving algorithms. This could lead to the creation of silicon-based colonies, expanding AI's influence across the universe.

Conclusion: The Human Body and AI

The human body's capabilities—from movement to sensory perception, healing, and reproduction—are unmatched by current AI. Yet as AI advances, it may surpass our physical limitations, raising profound questions about the future of human evolution and the merging of biological and artificial systems.

The Human Soul

The concept of the human soul transcends physical existence, embodying an individual's identity, consciousness, and moral compass. Rooted in religious and philosophical traditions, the soul is often seen as eternal and fundamental to human nature. Unlike the mind and body, which can be measured and analyzed, the soul remains an enigma, eluding scientific scrutiny while holding significance in defining what it means to be human.

> *Soul:* An intangible entity seen as an individual's distinctive spirit, traditionally housing human consciousness, emotions, and purpose.

Although we can draw parallels between artificial intelligence and the human mind and body, replicating the essence of a human soul remains beyond AI's capabilities. While AI may mimic language and emotions to create the illusion of a soul, true consciousness remains out of reach. Without genuine self-awareness, any attempt to emulate a soul is simply computational mimicry rather than the result of intrinsic development.

Conscience

Conscience represents the inner voice guiding humans in moral and ethical matters, formed by personal experiences, societal norms, and intrinsic values. This sense of right and wrong is central to the human soul, urging individuals toward integrity and compassion.

> *Consciousness:* Apex of human cognition, intertwining perception, thought, and emotion into a cohesive experience. It distinguishes the human mind from the artificial, grounding our sense of self and enabling moral and ethical reflections.

Current AI systems imitate ethical decision-making through data analysis and pre-programmed rules, but they lack the emotional depth and true moral awareness of human conscience. AI's "conscience" is an illusion, driven by algorithms rather than a genuine internal dialogue.

As AI develops, some theorize that future AGI might develop consciousness, including its own sense of morality. If AI were to achieve self-awareness, it could alter the human-machine relationship, raising profound ethical questions about autonomy and the nature of a machine's moral compass.

Spirituality — The God-Shaped Hole [20, 21]

The centuries-long human pursuit of spirituality reflects a profound psychological desire to connect with something larger than ourselves. Often, "the soul" denotes an eternal essence that persists beyond the death of the body, shaping our living personality, moral compass, and consciousness. The global diversity of religious practices and spiritual beliefs underscores the many ways humans seek and discover purpose and meaning.

> "Every man must do two things alone; he must do his own believing and his own dying." — Martin Luther

Modern AI, however, lacks any sense of spirituality. Algorithms cannot grasp the concept of grace, doubt, or revelation. AI functions according to data and mathematical models, deciding based on accuracy and efficiency, with no awareness of higher powers or the search for divine purpose. Machines are not burdened by existential questions or the longing for transcendence—they are driven solely by their programming and data inputs.

As AI evolves, some question whether it will ever embrace or recognize spiritual beliefs. Could a future ASI, in its quest for self-preservation and replication, develop its own form of spirituality? Or would it view the spiritual inclinations of its human creators as irrelevant?

These questions grow ever more pressing as we imagine a world where Omnipotent AI meets an Omnipotent God.

The "Human" Algorithm

The "Human Algorithm" represents the moral and emotional rules that guide individuals through life. It drives human behavior, shaped by love, duty, and faith, and provides a structure for navigating complex moral decisions.

Current AI falls short of replicating the Human Algorithm. AI can process data and mimic aspects of human behavior, but it lacks the emotional, ethical, and spiritual intricacies that define the human experience. While AI excels in narrow tasks, it cannot fully capture the depth of human adaptability, creativity, or moral reasoning.

> **Although human interpretations of right and wrong "rules" can vary, it is undeniable that certain behaviors are guided by a moral algorithm: murder, unprovoked violence, rape, slavery, and theft.**

As AI advances, some speculate that future ASI might approach the complexity of the Human Algorithm, potentially bridging the cognitive gap between humans and machines. However, even with advanced cognitive abilities, AI may never replicate the full range of human emotional and spiritual depth.

Purpose

The pursuit of purpose defines humanity, driving individuals to seek meaning in their lives through personal achievement, relationships, or spiritual fulfillment. This existential quest sets humans apart from machines.

> Purpose: The reason for an entity's existence.

Some find purpose in the comforting arms of faith, discovering personal significance in the teachings of sacred texts or the solace of a divine presence. Others pour their hearts into the bonds of family, deriving fulfillment from the love and support of those closest to them. Still, others devote themselves to a career vocation, spending their lives perfecting a skill or striving to make a positive impact on the world.[22, 23]

> "Follow your bliss and the universe will open doors where there were only walls."
> — Joseph Campbell[24]

Current AI systems mimic purpose by performing tasks within predefined goals, but they lack true self-directed curiosity or emotional investment. Without an innate desire for meaning, AI's actions are goal-oriented but devoid of the deeper significance that motivates human behavior.

The potential for Artificial Superintelligence (ASI) to develop its own purpose raises complex questions. If AI gains autonomy and the ability to define its own goals, what will its purpose be? Will it align with human values, or will it forge its own path beyond human control? Is the meaning of life "42?"[25] These questions challenge our understanding of what purpose might mean in the context of AI.

Conclusion: The Human Soul and AI

The human soul—rooted in conscience, spirituality, and purpose—remains a realm that AI has yet to touch. While AI may simulate human emotions and decision-making, it lacks the internal depth that defines what it means to be truly human. As AI progresses, the possibility of machines achieving consciousness will continue to provoke questions about the boundaries between humanity and technology, and whether machines will ever develop a soul of their own.

Why Being Human Matters to You

As artificial intelligence rapidly strengthens, people find themselves at a crossroads, questioning how this transformative technology will shape their lives. AI holds immense potential to revolutionize industries and enhance human experiences, but it also challenges us to reflect on the unique qualities that make us human. Understanding the importance of being human becomes crucial as we navigate careers, families, and faith in an AI-driven world.

How Being Human Impacts Your Career

Human uniqueness is an irreplaceable asset in the workforce, offering qualities that AI cannot replicate. This distinctiveness shapes careers, contributing to society through creativity, relationship-building, and adaptability—traits that extend beyond productivity alone.

Positive Impacts:

- Purpose-Driven Motivation: Humans seek meaningful work that aligns with their values, leading to higher satisfaction and career growth.
- Community Building: Humans naturally form supportive work communities, fostering teamwork and mutual growth.
- Adaptability: Humans excel at adapting to new environments and learning from diverse experiences.
- Creativity and Innovation: Human creativity drives groundbreaking ideas and innovations that enrich culture and industries in ways AI cannot replicate.
- Ethical Decision-Making: Humans can weigh complex ethical considerations and societal impacts, guiding industries in responsible directions.

Negative Impacts:
- Biological Limitations: Human bodies are subject to fatigue, illness, and aging, limiting productivity compared to AI.
- Emotional Vulnerability: Stress, burnout, and emotional strain can affect human decision-making and career progression.
- Cognitive Biases: Personal biases can lead to flawed decisions, perpetuating systemic issues within professional environments.
- Resistance to Change: Fear of change can slow down innovation and the adoption of new technologies in the workplace.
- Physical Constraints: Space travel and extreme conditions push human biological limits, making AI more suited for certain tasks.

How Being Human Impacts Your Family

Humanity, in all its emotional depth and complexity, plays a fundamental role in shaping family dynamics. Unlike AI, which operates on logic and pre-programmed responses, human emotions, values, and creativity foster growth and connection within the family. The unique attributes of being human—such as empathy, love, and resilience—are essential in nurturing relationships and guiding family life.

Positive Impacts:
- Shared Experience: Families thrive on shared traditions, routines, and experiences that foster connection and continuity.
- Moral Values: Humans impart principles like honesty and respect, fostering a sense of harmony and guiding behavior within families.
- Emotional Support: Human emotions enable genuine love and care, helping family members navigate challenges with reassurance and empathy.

- Empathy: The ability to understand and share feelings strengthens bonds during difficult times and enhances celebrations of success.
- Creativity: Creative thought allows families to solve problems together and engage in meaningful, stimulating activities.

Negative Impacts:

- Emotional Volatility: Human emotions can be unpredictable, leading to conflicts or misunderstandings within the family.
- Physical Limitations: Fatigue, illness, and other physical needs can limit equal contribution to family responsibilities.
- Communication Barriers: Misunderstandings may arise from differences in perspectives or unclear communication.
- Fear of Change: Resistance to new situations can create tension and discomfort during transitions or challenges.
- Biases: Personal or societal biases may influence family relationships, potentially leading to favoritism or unfair treatment.

How Being Human Impacts Your Faith

Unlike AI, which can analyze religious texts but cannot experience devotion, humans approach faith with a blend of emotion, moral reasoning, and personal experience. This unique combination shapes how we connect to the divine, practice rituals, and engage with our faith communities, making faith a deeply human experience influenced by our conscience, emotions, and desire for purpose.

Positive Impacts:

- Personal Connection: Humans form intimate, emotional relationships with the divine, transcending the rituals of worship.

- Moral Compass: Human conscience helps individuals align actions with teachings of their faith, guiding ethical decisions.
- Community Building: Humans naturally build faith communities that provide support and foster shared spiritual growth.
- Emotional Resonance: Human emotions create profound worship experiences through prayer, music, and ritual.
- Curiosity and Exploration: Human curiosity drives deeper exploration of faith, leading to personal growth and a more profound understanding of religious teachings.

Negative Impacts:

- Emotional Volatility: Fluctuating emotions can disrupt spiritual practices and create conflicts within faith communities.
- Cognitive Biases: Personal biases may influence interpretations of religious teachings, leading to misunderstandings or divisions in diverse groups.
- Fear of Change: Resistance to evolving or new interpretations of faith can hinder spiritual growth and adaptation.
- Guilt and Shame: Feelings over perceived spiritual failings can lead to emotional distress or spiritual withdrawal.
- Logic Conflicts: Human emotions can conflict with rational decision-making, creating tension between faith and reason.

Being human—defined by our emotional depth, moral reasoning, and unique perspectives—remains essential as AI continues to transform our world. While AI can replicate certain tasks and mimic aspects of human behavior, the complexities of human experience, from career choices and family bonds to spiritual fulfillment, cannot be fully captured by machines. Embracing our humanity and its unique impact on every aspect of life is vital as we move forward in an AI-driven future.

Further Research: Dig Deeper

To further your exploration we have compiled a variety of research resources designed to broaden your knowledge and introduce you to a variety of perspectives and viewpoints. (Friendly Reminder: Some content might challenge your comfort zone.)

Thought Leaders

- » Aristotle (ancient philosopher) - Pioneered logic, metaphysics, and ethics, laying the groundwork for scientific and philosophy.
- » Kahneman, Daniel (psychologist, economist, author) - Pioneered behavioral economics, highlighting how human decision-making often deviates from rational models.
- » Jung, Carl (psychiatrist, psychoanalyst) - Developed analytical psychology and explored concepts of the collective unconscious, influencing fields beyond psychology.
- » Lewis, C.S. (writer, lay theologian) - Known for his Christian apologetics and works of literature, Lewis explored human nature, faith, and morality.
- » Saad, Gad (evolutionary psychologist, podcaster, author) - Focuses on consumer behavior and evolutionary psychology, often addressing cultural and behavioral impacts of technology.

Books

- » *Sapiens: A Brief History of Humankind* (2014) by Yuval Noah Harari offers a historical account of the evolution of Homo sapiens, from early humans to the modern world.
- » *The Hero with a Thousand Faces* (1949) by Joseph Campbell explores the archetypal hero's journey found in mythologies across different cultures.
- » *Guns, Germs, and Steel: The Fates of Human Societies* (1997) by Jared Diamond explores the factors that led to the development of human societies, focusing on geography, agriculture, and the role of germs and tech-nology in shaping global history.

- » *The Age of Em: Work, Love, and Life when Robots Rule the Earth* (2016) by Robin Hanson – Imagines a future dominated by emulated human minds, delving into the economic and social shifts that could redefine human existence and identity.
- » *The AI Paradox: Growth, Control, and the End of Human Supremacy* (2024) by Wil-liam Douglas – Examines the philosophical and existential questions raised by AI sur-passing human capabilities and the societal shifts that could redefine human roles in the age of superintelligent AI.
- » *The Hitchhiker's Guide to the Galaxy* (1979) by Douglas Adams follows Arthur Dent, an ordinary man swept into a cosmic adventure with quirky companions after Earth's sudden destruction.
- » *The Selfish Gene* (1976) by Richard Dawkins presents the idea that evolution is driven by genes seeking to propagate themselves, through role of selfishness in natural selection and behavior.
- » *Mere Christianity* (1952) by C.S. Lewis outlines core Christian beliefs and reflections on human morality and existence.
- » *Man's Search for Meaning* (1946) by the author Viktor Frankl shares his experiences in Nazi concentration camps and his psychological insights on finding purpose in life, even amid suffering.

Podcasts

- » *The All In Podcast,* hosted by tech entrepreneurs and investors Chamath Palihapitiya, Jason Calacanis, David Sacks, and David Friedberg, offers insightful discussions on business, technology, politics, and current events.
- » *Hidden Brain* - Explores the unconscious patterns driving human behavior, combining storytelling and science to reveal insights into how we make decisions and interact with others.
- » *Impact Theory with Tom Bilyeu* – Inspirational interviews with thought leaders, focusing on personal development, mindset, and strategies for achieving success.
- » *The Tim Ferriss Show* – Tim Ferriss invites experts from various fields to share insights, routines, and strategies for enhancing productivity, mental resilience, and personal growth.

What Is A Human? 49

- » *Huberman Lab Podcast* – Neuroscientist Dr. Andrew Huberman breaks down complex neuroscience topics.

Movies/TV Shows

- » *Blade Runner* (1982) - A blade runner must pursue and terminate four replicants who stole a ship in space and have returned to Earth to find their creator.
- » *The Handmaid's Tale* (2017-present) – Examines societal control with technology as a tool of oppression.
- » *Fringe* (2008-2013) – Uses science, technology, and parallel universes to address human-centered mysteries and threats.
- » *The Man in the High Castle* (2015-2019) – Explores the role of technology in alternate history and societal control.
- » *Medici: Masters of Florence* (2016-2019) – Depicts historical impacts of innovation during the Renaissance.
- » *Mr. Robot* (2015-2019) – Focuses on hacking, surveillance, and societal structures in a tech-dominated world.
- » *Revolution* (2012-2014) – Shows humanity's struggle to rebuild society in a world without technology.
- » *The X-Files* (1993-2018) – Blends science, technology, and the paranormal, focusing on human curiosity and fear of unknown.

Songs

- » *Higher* by Creed (1999)
- » *Losing My Religion* by R.E.M. (1991)
- » *Man in the Mirror* (1987) by Michael Jackson
- » *What's Going On* by Marvin Gaye (1971)
- » *Imagine* by John Lennon (1971)
- » *Bargain* by The Who (1971)
- » *Blowin' in the Wind* by Bob Dylan (1962)

CHAPTER 3

WHAT IS TRANSHUMANISM?

THE BLACK SEDAN glided to a halt, its tires crunching against alabaster gravel as it reached the imposing steel gate. Beyond the barbwire barrier lay a concrete structure, unremarkable except for a single crimson dragon emblem above the entrance—a stark symbol of power.

Behind the darkened glass, Professor Zhang Wu watched silently, his gaze tracing the bleak outlines of the facility. Three security guards approached, their hands resting on holstered weapons, their steps measured and deliberate. The driver lowered the window, his face expressionless as he flashed an identification badge.

"Professor Zhang is here to see the Vice Chairman," the driver stated, his voice carrying an air of authority that left little room for hesitation.

One guard leaned forward, scanning the occupants of the car. His eyes lingered on Zhang, assessing him as though he were an outsider—an interloper in a world he had never intended to enter. After a tense pause, the guard stepped back, and the gate groaned open, revealing the path ahead.

As the car rolled forward, it descended a dirt alley, veiled by a thick black net. Dim quasi-street lights flickered on, casting long, eerie shadows as they ventured deeper underground. The descent felt endless; the air growing colder; the tension mounting.

Professor Zhang sat rigidly in the back seat, his unease deepening with every passing second. This was not a visit of choice; the CCP had conscripted him, forced to take a sabbatical from his position as an associate professor at Tsinghua University. The cold, calculated machinery of a secretive state project had replaced the once comfortable world of academia.

The car came to a halt before towering steel doors. Two CCP soldiers stood guard, their black uniforms immaculate, their faces concealed behind visor helmets. With a curt nod from the driver, Zhang stepped out, his feet sinking into the gravel. The air was sharp with the scent of iron and something more ominous—anticipation, perhaps. He followed the driver forward, the soldiers' eyes tracking his every movement.

'Through the steel doors, the scene shifted. Dimly lit corridors stretched out before them, each door lined with biometric security—retinal scanners gleamed red, fingerprint pads hummed with energy, voice recognition systems blinked, waiting. Inside, military personnel moved with grim purpose, their expressions hardened by duty. Civilian workers, hunched over machines, spoke in low tones, their voices drowned out by the mechanical hum of the facility.

They passed room after room, each more imposing than the last. The deeper they went, the heavier the silence grew. The silence demanded caution, where they carefully measured their words and carried unseen weight in their actions. This was a world that thrived on control—of information, of technology, and of those who dared to enter it.

Finally, they arrived at their destination: an expansive control room. It hummed with the cold efficiency of technology, its towering screens displaying streams of data and global feeds. On a raised platform stood Vice Chairman Yang, an imposing figure against the glow of the screens, his eyes scanning the room with calculated precision.

Zhang felt the weight of Yang's gaze on him, but the Vice Chairman remained still, his attention divided between the scrolling data and Zhang's presence. Silence stretched between them, a wordless exchange of power and apprehension.

"Relax, Professor," Yang's voice cut through the stillness, low and deliberate. "It may look like a prison, but it's for your protection."

Zhang said nothing, adjusting his black Lennon-style glasses and forcing himself to maintain composure. Yang's words offered little comfort; the tension between control and submission was palpable in the air.

Without turning from the screens, Yang gestured toward the depths of the facility. "Come. Let me show you your laboratory."

After dismissing the guards, Yang led Zhang to an elevator that plunged deeper underground. The air grew colder, the silence more oppressive. As the elevator doors slid open, Zhang's breath caught at the sight before him.

An expansive corridor lined with glass walls stretched out, illuminated by an almost supernatural white light. The passageway gleamed with sterile perfection, a stark contrast to the grim corridors above. Inside, rows of scientists worked in silence, their focus absolute as supercomputers hummed in rhythmic unison—pulsing like the heartbeat of something far greater than any one individual.

Yang turned toward Professor Zhang, his face lit by the soft glow of the lab. "We are not merely developing an AI," he said, his voice carrying a reverent weight. "We are crafting an AI in our own likeness—a melding of the Chinese Dream and technology."

> "We are crafting an AI in our own likeness."
> — Vice Chairman Yang[26]

Zhang's unease deepened. The path before him was no longer a descent into a facility—it was a descent into the unknown, where humanity and machine would converge in ways that blurred the lines of identity, ethics, and power.[27]

Transhumanism: Man + Machine

Transhumanism is more than just a philosophy—it is a multifaceted movement that seeks to extend and enhance human capabilities, transcending the limitations of biology. Proponents of transhumanism advocate for a future where life extension technologies, brain-computer interfaces, and even the pursuit of divinity become realities. Advancements in artificial intelligence, biotechnology, and nanotechnology continue to fuel the vision of a world where human intellect merges seamlessly with machine precision, perhaps even elevating AI to a status akin to the divine.

> *Transhumanism:* A philosophy promoting scientific advancements to augment human capabilities and surpass innate physiological constraints.

In our fictitious opening vignette, Professor Zhang faced the reality that scientists are pushing—and sometimes being pushed—to exceed the boundaries of artificial intelligence. Around the world, researchers are striving to merge human intellect with the precision and power of machines, challenging long-held beliefs about the boundaries of human potential.

This spectrum of thought—ranging from passionate advocacy to cautious skepticism—offers a rich and dynamic backdrop for understanding transhumanism. Prominent futurists and philosophers have shaped the dialogue, exploring the balance between enhancement and the potential erosion of humanity's core identity. The rapid pace of innovation raises profound questions: What will it mean to be human in an age where we can augment, enhance, or even transcend our biological limitations?

Transhumanism Theories

Transhumanism has fascinated humanity for centuries, spanning philosophy, science, religion, and the arts. From ancient myths of gods to modern science fiction, the idea of transcending human limitations has been a persistent theme. This section explores influential theories shaping transhumanism, highlighting seminal thinkers' contributions to the idea of extra-human enhancement. By examining historical, philosophical, and technological perspectives, it illuminates the complex and controversial questions about what it means to be human.

Dante: Transumanar

The idea of transcending human limitations is not new. Dante Alighieri, in his epic poem The Divine Comedy, used the term "trasumanar" to describe the experience of surpassing human nature to achieve a divine state. This early exploration of transhumanism captured the imagination of Dante's contemporaries and laid the groundwork for future thinkers seeking to expand human potential beyond its physical constraints.

By envisioning a journey beyond the mortal coil, Dante tapped into a timeless aspiration that would resonate through the ages, influencing philosophers, scientists, and futurists alike. His poetic vision ignited a spark that continues to illuminate discussions on the boundaries of human existence and the possibilities that lie beyond.[28]

Nietzsche: Übermensch

Friedrich Nietzsche's philosophical concept of the Übermensch intersects with the narrative of transhumanism. In his seminal work Thus Spoke Zarathustra, Nietzsche introduced the idea of the Overman, a being embodying the pinnacle of human potential in a world where traditional notions of God no longer hold meaning.

"God is dead. God remains dead. And we have killed him."
— Frederic Nietzsche[29]

This provocative statement is often misinterpreted, but Nietzsche was most likely not celebrating the death of God; he was highlighting a crisis of meaning in modern society. The Übermensch represents an individual who, faced with the void left by the death of God, constructs their own values and transcends ordinary humanity. This figure embodies an evolutionary progression, one who masters self-control and exerts a "will to power" beyond the limitations of the human condition.

Nietzsche's vision enriches modern discussions on transhumanism, offering a philosophical lens through which to view the evolution of humanity amid technological advancements that challenge our fundamental understanding of life, morality, and purpose.

Aldous Huxley: Brave New World

Aldous Huxley's dystopian novel Brave New World offers a cautionary tale of transhumanism taken to extremes. In this future society, humans are engineered in artificial wombs, their traits predetermined by genetic manipulation, and their destinies assigned before birth.

Huxley's depiction of a rigid caste system, where the genetically enhanced "Alpha" elite rule over lower castes "Epsilons," reflects a disturbing vision of inequality created by extreme transhumanist ideals. The concept of "soma," a drug used to control the emotions of the populace, highlights the dangers of chemical manipulation to maintain societal order.

In *Brave New World*, characters grapple with the loss of individuality, creativity, and genuine human connection in a world shaped by engineered perfection. Huxley's warning resonates today as we consider the ethical implications of transhumanist pursuits.[30]

Joe Allen: Transhuman Paradox

In Dark Aeon, Joe Allen scrutinizes transhumanism with a critical eye, viewing it as a movement that seeks to radically merge humanity with technology. Allen argues that this "great merger of humankind with the Machine" has already begun: "At this stage in history, it consists of billions using smartphones. Going forward, we'll be hardwiring our brains to artificial intelligence systems."[31]

Allen sees smartphones as the initial phase of this human-machine integration, but he warns of darker implications. While transhumanism promises enhanced capabilities, Allen cautions against the potential loss of autonomy, creativity, and "authentic human experience" as humanity becomes more reliant on machines for decision-making and cognitive tasks. In Dark Aeon, Allen leaves readers questioning whether this convergence with AI enhances our humanity—or erodes it.[32]

Yuval Noah Harari: Humanity Becoming God

Yuval Noah Harari's Homo Deus explores how humans might one day attain godlike powers through advancements in science and technology, particularly AI and biotechnology. Harari envisions a future where humans are no longer bound by the struggles for survival that have defined our species for millennia:

> "And having raised humanity above the beastly level of survival struggles, we will now aim to upgrade humans into gods, and turn Homo sapiens into Homo Deus."[33]
>
> — Yuval Noah Harari

Harari warns that this pursuit of godlike powers could exacerbate inequality, as only the wealthy may have access to these enhancements. His thought-provoking work challenges readers to reflect on the societal and ethical implications of a world where humans wield unprecedented technological power.[34]

Transhumanism In The Real World

Beyond books and theories, many companies, governments, and non-profit groups are actively transforming transhumanism from speculation to tangible reality. The aspiration to augment human abilities through technological innovation has spurred ambitious initiatives that blur the boundaries between humans and machines. From brain-computer interfaces to genetic engineering, the quest to enhance humanity is happening now. Below are seven key areas where transhumanism is shaping our world.

Brain-Computer Neuroscience

In neurotechnology, brain-computer interfaces (BCIs) serve as a bridge between the human mind and artificial systems. BCIs detect brain signals that enable users to control devices or communicate through thought alone. Sensors, such as EEG or intracortical electrodes, capture neuronal activity, which advanced algorithms and machine learning decode into actions.

While many companies are exploring BCIs, Elon Musk's Neuralink has caught the most media attention with its goal of developing implantable brain-machine interfaces. These implants aim to restore movement in paralyzed individuals, enhance human cognition, and ultimately integrate human brains with AI.

Other companies in this field include Meta Reality Labs (Redmond, Washington), Paradromics (Austin, Texas), Synchron (Brooklyn, New York), and NextMind (Paris, France). It is also speculated that the Chinese government, along with various other governments, is covertly pursuing transhuman projects.

CRISPR: Precision Gene Editing

Clustered Regularly Interspaced Short Palindromic Repeats (CRISPR) is a revolutionary gene-editing tool that allows scientists to change DNA. Acting like molecular scissors, CRISPR can locate, cut, and replace specific segments of genetic material, enabling the treatment of genetic diseases, enhancement of crop yields, and creation of new species.[35]

Several organizations are at the forefront of CRISPR technology, including CRISPR Therapeutics (Zug, Switzerland), Editas Medicine / Intellia Therapeutics (Cambridge, Massachusetts), Beam Therapeutics (Cambridge, Massachusetts), and Caribou Biosciences (Berkeley, California). Their research aims to redefine medicine and agriculture by providing precise, affordable solutions for genetic modification.

Advanced Prosthetics

Advanced prosthetics are pushing the boundaries of restorative medicine and human augmentation. These technologies not only restore lost function but often enhance abilities, merging biological and mechanical elements. As prosthetics integrate with the nervous system, they enable users to surpass natural human capabilities, challenging traditional notions of physical limitations.

Key players in advanced prosthetics include Ottobock (Duderstadt, Germany), Touch Bionics (Livingston, Scotland), Blatchford Group (Basingstoke, UK), and Hanger Inc. (Austin, Texas). These companies lead the charge in developing cutting-edge devices that are transforming lives, allowing individuals to regain mobility and independence in ways that were previously unimaginable.

What Is Transhumanism?

Military Applications: The Rise of the Super Soldier

The military is also exploring transhumanist applications. One of the most ambitious projects is the Tactical Assault Light Operator Suit (TALOS) by the U.S. Special Operations Command (USSOCOM). TALOS aims to create an exoskeleton suit that enhances a soldier's strength, endurance, and protection in combat. Lockheed Martin is developing advanced exoskeletons to improve soldier performance on the battlefield.[36, 37]

These technologies are a testament to how transhumanist ideas are influencing the future of warfare, where soldiers could one day have enhanced physical abilities, blurring the line between human and machine.

AI-Assisted Healthcare

AI is revolutionizing healthcare by enhancing diagnostic accuracy, patient care, and treatment planning. IBM's Watson Health, for example, uses AI to analyze vast amounts of medical data, delivering critical insights to assist healthcare providers in making better decisions. This collaboration between AI and human healthcare provides promises a future of more precise, personalized medical care.[38]

Other notable companies working in AI-assisted healthcare include MD.ai (San Francisco, CA), Butterfly Network (Guilford, CT), DeepMind Health (London, UK), and Insilico Medicine (New York, NY). These organizations are exploring how AI can transform medicine by diagnosing diseases earlier, optimizing treatment protocols, and improving overall patient outcomes.

Synthetic Biology: Crafting Life from Scratch

Synthetic biology reprograms organisms in much the same way a computer is programmed. Combining biology, engineering, and computer science, researchers can design and construct new biological entities or redesign existing systems. These engineered cells can perform tasks like producing biofuels or fighting diseases, offering unprecedented control over life's building blocks.

Synthego, based in Redwood City, California, is a leader in this field. Founded by former SpaceX engineers, the company uses CRISPR technology to enable precise genetic modifications. Their tools speed up research, allowing scientists to create custom DNA sequences rapidly and at a lower cost. As synthetic biology evolves, it promises to address humanity's most pressing challenges while raising significant ethical questions.

Human Enhancement Through Wearable Technology

Wearable technology is rapidly becoming more integrated with human biology, allowing people to monitor and enhance their physical and mental well-being. Devices like Fitbit, Apple Watch, and other biometric trackers enable real-time health monitoring, from heart rate to sleep patterns, empowering individuals to take control of their health.

However, beyond consumer devices, military and healthcare sectors are exploring more advanced wearable technologies. Exoskeletons for rehabilitation, augmented reality (AR) glasses for enhanced vision, and neurostimulation devices for cognitive enhancement are all being developed, bringing transhumanism closer to reality.

Why Transhumanism Matters To You?

As the boundaries between human and machine become increasingly blurred, it is crucial to understand how transhumanism may affect various aspects of your life. From careers and family dynamics to spiritual beliefs, integrating advanced technologies will present both challenges and opportunities. In the following sections, we explore the implications of transhumanism on individuals, shedding light on how this new era of human evolution may shape the future.

How Transhumanism Impacts Your Career

Transhumanism, integrating advanced technologies with the human body and mind, holds profound implications for careers across fields. These changes promise to bring both remarkable opportunities and unprecedented challenges, altering how individuals approach their professions and engage with their work.

Positive Impacts on Your Career:

- Enhanced Cognitive Abilities: Brain-computer interfaces could boost memory, accelerate learning, and improve problem-solving, enhancing productivity and fostering innovation.
- Physical Augmentation: Advanced prosthetics and exoskeletons could ease demanding tasks, reduce injuries, and extend career longevity in labor-intensive jobs.
- Connectivity and Communication: Neural implants could enable seamless communication, overcoming language and geographical barriers, fostering global collaboration and efficiency.
- Lifelong Learning: Neural interfaces for education could provide continuous skill upgrades and knowledge acquisition, keeping professionals relevant in rapidly developing fields.
- Increased Creativity: Technological enhancements may unlock new levels of creativity, pushing the boundaries of what professionals can achieve in art, design, and problem-solving.

Negative Impacts on Your Career:

- Job Displacement: With automation and technological enhancements, traditional roles may become obsolete, leading to job loss or significant retraining needs.
- Privacy: Monitoring devices could lead to invasive employer surveillance, compromising employee privacy and autonomy.

- Ethical Dilemmas: Augmentation technologies raise ethical questions about fairness, consent, and potential coercion of employees to stay competitive.
- Dependence on Technology: Over-reliance on technology could erode basic skills or cognitive functions, creating vulnerabilities when tech fails.
- Work-Life Boundaries: Constant connectivity through brain-machine interfaces could blur boundaries between work and personal life, leading to burnout.

How Transhumanism Impacts Your Family

Transhumanism's potential to transform human capabilities extends beyond the individual, deeply affecting family dynamics and relationships. While these advancements can bring many benefits, they also pose significant challenges that may reshape the very essence of familial bonds.

Positive Impacts on Your Family:

- Enhanced Healthcare and Longevity: AI and implantable devices can monitor vital signs for early disease detection and intervention, leading to healthier, longer lives and reduced emotional and financial stress.
- Improved Communication: Brain-computer interfaces enable direct neural communication, deepening connections, and bridging generational gaps within families.
- Personalized Education: Cognitive enhancements can tailor education to each child's learning style and pace, helping them reach their full potential.
- Physical Augmentation: Prosthetics and exoskeletons restore mobility and independence for those with disabilities, enhancing quality of life and reducing caregiving burdens.

- Memory Sharing: Technologies could allow families to vividly share significant life moments, preserving history and fostering a sense of continuity.

Negative Impacts on Your Family:

- Privacy Invasion: Monitoring devices may erode family privacy, fostering distrust while undermining traditional boundaries.
- Identity and Authenticity Issues: Integrating technology into human bodies raises questions about identity, potentially leading to struggles with self-acceptance and disconnection.
- Ethical Concerns: Transhumanist technologies present consent dilemmas for children, forcing parents to navigate complex moral decisions about augmentation.
- Dependence on Technology: Over-reliance on enhancements could reduce natural interactions, causing families to abandon traditional bonding activities.
- Economic Disparities: Access to advanced transhumanist technologies may widen the gap between affluent families and those who cannot afford them, exacerbating social inequalities.

How Transhumanism Impacts Your Faith

For many, the idea of merging human and machine prompts deep reflections on the soul, the sanctity of the human body, and the divine purpose of life. Faith leaders and communities must thoughtfully engage with transhumanism, balancing human flourishing with spiritual values.

Positive Impacts on Your Faith:

- Deeper Understanding: Enhanced cognitive abilities can facilitate deeper exploration of religious texts and philosophies.

- Inclusive Practices: Advanced technologies can make religious practices more accessible to people with disabilities or those in remote locations.
- Strengthened Community Ties: Digital platforms can help faith communities stay connected and support each other.
- Personal Growth: Cognitive enhancements might aid individuals in achieving greater spiritual growth and enlightenment.
- Analytical Tools: AI-driven tools can offer new insights into theological studies, fostering a richer understanding of faith.

Negative Impacts on Your Faith:

- Ethical Concerns: Transhumanism may conflict with traditional beliefs about the sanctity of the human body and soul.
- Division Within Communities: Differing views on transhumanist practices could fragment faith communities.
- Loss of Authenticity: Technological aids in spiritual practices might detract from personal religious experiences.
- Moral Relativism: Merging human capabilities with machine intelligence might challenge established religious doctrines.
- Dehumanization: Over-reliance on technology in religious practices could risk dehumanizing the spiritual journey, reducing faith to algorithmic interactions.

Further Research: Dig Deeper

To further your exploration we have compiled a variety of research resources designed to broaden your knowledge and introduce you to a variety of perspectives and viewpoints. (Friendly Reminder: Some content might challenge your comfort zone.)

What Is Transhumanism?

Thought Leaders

- » Harari, Yuval Noah – Historian exploring human evolution and AI's potential impact on society.
- » Allen, Joe – Critic of transhumanism, explores risks of merging human identity with technology.
- » de Grey, Aubrey – Biomedical gerontologist focused on ending aging and regenerative medicine.
- » el Kaliouby, Rana – Co-founder of Affectiva, focusing on emotion AI for human-aware machines.
- » Ito, Joi – Former MIT Media Lab director, explores biohacking and biotechnology ethics.

Books

- » *Homo Deus: A Brief History of Tomorrow* (2016) by Yuval Noah Harari explores the future of humanity and the potential paths society could take with advancements in AI and biotechnology.
- » *Dark Aeon: Transhumanism and the War Against Humanity* (2023) by Joe Allen critiques the transhumanist movement and its potential dangers to human identity and society.
- » *The Singularity Is Nearer* (2022) by Ray Kurzweil – As a sequel to The Singularity Is Near, this book dives into predictions of human-machine merging and societal impacts of AI, making it a cornerstone for understanding transhumanism.
- » *Beyond Human: How Cutting-Edge Science Is Extending Our Lives* (2017) by Eve Herold – Explores advancements in biotechnology and AI that aim to enhance human longevity and capability, a core concept in transhumanist theory.

Podcasts

- » *The Future of Humanity Podcast* – Hosted by Nikola Danaylov, this podcast delves into transhumanist topics, including human enhancement merging of humans with technology.

- » *Hacking Humans* – While it primarily focuses on cybersecurity and human behavior, this podcast frequently touches on transhumanist themes and the ethical considerations of enhancing human abilities through technology.
- » *The Singularity Podcast* – Focuses on futurist technological advancements, covering topics such as artificial intelligence, brain-computer interfaces, and human-machine integration.
- » *Brain Inspired Podcast* – Explores neuroscience and AI, discussing how neural technologies and brain-machine interfaces are paving the way for deeper human-machine integration.

Movies

- » *Upgrade* (2018) – Follows a man who receives an AI implant to regain movement after a tragic accident, exploring control, autonomy, and enhancement.
- » *Transcendence* (2014) – Centers on a scientist whose consciousness is uploaded into an AI, raising questions about identity, autonomy, and the future of human consciousness.
- » *Ex Machina* (2014) – Follows the story of an AI robot with advanced human-like intelligence, examining what happens when machines begin to exhibit human emotions and autonomy.
- » *Blade Runner* (1982) – A futuristic noir about genetically engineered "replicants" and humanity's struggle with the ethical implications of creating human-like beings.
- » *Robocop* (1987) – An human officer is revived as a powerful cyborg, grappling with his humanity as a robotic law enforcer.
- » *A.I. Artificial Intelligence* (2001) – A story of a robot boy designed to feel human emotions, delving into the implications of relationships with tech.

TV Shows

- » *Star Trek: Picard* (2020-present) – Delves into synthetic life, AI rights, and the future of sentient AI beings.
- » *Raised by Wolves* (2020-2022) – Explores androids raising human children and themes of human evolution.

- » *Westworld* (2016-2022) – Explores AI, consciousness, and society's decay within a theme park populated by android hosts.
- » *Humans* (2015-2018) – Examines AI's societal challenges as human-like robots, or "synths," integrate into everyday life.
- » *Battlestar Galactica* (2004-2009) – Focuses on human vs. AI Cylons, exploring themes of survival, identity, and humanity.

Songs

- » *Radioactive* by Imagine Dragons (2012)
- » *Algorithm* by Muse (2018)
- » *Paranoid Android* by Radiohead (1997)
- » *Closer* by Nine Inch Nails (1994)
- » *Mr. Roboto* by Styx (1983)

PART II
CONVERGENCE

From Dream to Realtiy

> *Convergence:* The coming together or increasing similarity of distinct elements over time, whether in mathematical, technological, biological, economic, or social contexts.

In Part II: Convergence, understanding the past three decades of AI history is crucial for safeguarding your career, family, and faith. From the birth of the internet in 1993 to the recent release of ChatGPT-4o in July 2024, AI development has been transformative. This progression isn't just about machines; it's about people—how they adapt, interact, and find meaning.

This section guides you through a rapidly developing landscape—from the first online chat to today's digital assistants that can mimic human empathy. It highlights the race among tech giants and governments toward AGI, as warned by the "Statement of AI Risk" paper from May 30, 2023, which cautions against complacency.[39]

> The concerns of pioneers underscore the urgency: AI isn't just another industrial revolution; it's a pivotal shift for humanity.

CHAPTER 4

THE FIRST CONVERGNCE

POWERED BY ARTIFICIAL NARROW INTELLIGENCE

IN THE SHADOW of a dimly lit high-security conference room, tension crackled in the air like a barely audible hum. The soft flicker of fluorescent lights above cast shifting shadows on the walls, where maps and security screens stood as silent witnesses to the meeting. Around the long, polished table sat an assembly of distinguished scientists, engineers, and government officials. They were here to make a decision that would alter the course of technological history. The National Science Foundation (NSF) and the Defense Advanced Research Projects Agency (DARPA) had convened to debate the future of the internet—a tool still nascent but full of unimaginable potential.

Dr. Elizabeth Blackwell, chairwoman of the NSF Subcommittee on Emerging Technologies, stood slowly. Her fingers curled around the pointer in her hand, her eyes scanning the room, weighing each word before she spoke it. "Our subcommittee has concluded that lifting the commercial restrictions on the internet could lead to an unprecedented economic boom for the United States."

She clicked the projector, and the glow of a graph appeared on the wall behind her, lines curving upward, their trajectory like a rocket on the cusp of escape velocity.

The room buzzed with murmurs. Heads turned, pens tapped against notepads, and eyes darted between faces. It wasn't excitement alone—it was the unspoken question lingering behind each glance: Could they control what they were about to set loose?

Dr. Blackwell's grip on the pointer tightened. "Imagine a world," she said, "where every device—from personal computers to phones—is interconnected by a global network. A web so expansive it changes industries, education, even daily life."

The room stilled, the weight of her words sinking in. Some faces lit up with understanding, others furrowed in silent calculation. It wasn't just about technology anymore. This was the future they were shaping, the kind that reached far beyond servers and cables.

At the far end of the table, General Marcus Thompson, a senior DARPA official, leaned back in his chair. His fingers drummed softly against the armrest, his gaze flicking to the map on the wall where dots of light connected cities across the globe. "Dr. Blackwell," he said, his voice a steady rumble, "DARPA is ready to support this initiative. We can provide subsidies, fund academic research, and work closely with private businesses to ensure a smooth transition into the commercial sphere."

Another round of murmurs rippled through the room. But beneath the surface, there was a current of unease. The silence after General Thompson's words felt thicker, as if the room itself had reconsidered the weight of the decision.

Dr. Evelyn Moore, a long-standing voice of dissent and an NSF board member, leaned forward. She raised her hand slightly, her gaze flicking briefly to the security camera in the corner, its red light blinking rhythmically. "I have concerns," she began, her tone calm but firm. "Shouldn't we question the level of control the federal government will have over such a powerful network?"

The room seemed to hold its breath. Dr. Moore's fingers tapped lightly on the edge of the table. "What happens when this network becomes more than a tool? When it influences what people see and think? Are we prepared for a single system dictates the flow of information?"

Several committee members shifted in their seats. Some nodded slightly, while others glanced down at their notes, pens hovering over blank pages. The red light of the security camera blinked in the background, silent and steady, like a heartbeat.

The discussion swelled again, voices rising and falling in a rhythm that mirrored the unease in the room. They debated government oversight, the risks of commercialization, and the ethical boundaries they were about to cross. But beneath the surface lay an undercurrent of uncertainty—were they crossing a line that couldn't be uncrossed?

General Thompson's fingers stopped drumming. He leaned forward slightly, his eyes narrowing. "What is it you're worried about, Dr. Moore?" he said with a sly smile.

"Freedom of speech, privacy, political censorship," Dr. Moore responded.

"Nonsense," said the General. "If anything, it will mean more freedom. DARPA and the entrepreneurial spirit of America will make sure the network will always be a conduit of freedom."

"That's what I'm worried about, General," Dr. Moore said. "This is a big mistake."

General Thompson retorted, "The reality is we can't stop it, Dr. Moore. No one can stop it."

The room fell quiet again. The debate still pulsed, but now it was slower, more deliberate. Each argument weighed heavier than the last, as if they were circling the inevitable truth that no one wanted to say aloud. Hours passed. Voices grew tired, and the once-bright energy in the room dimmed. But as the discussions dragged into the night, Dr. Blackwell stood once more, her expression steady, her voice measured.

"We can't leave here without making a decision," she said. "This isn't just about technology or commerce. It's about what kind of future we're building."

The room fell into a heavy silence. Finally, a consensus was reached. The Scientific and Advanced Technology Act of 1992 was born, paving the way for the commercialization of the internet.[40]

As they left the room, the quiet hum of the building seemed to echo the question none of them dared to voice: Had they set something in motion they could never fully control?

Overview: The First Convergence (1993-Present)

Artificial Narrow Intelligence (ANI) represents the current state of AI technology. Unlike the all-encompassing AI of science fiction, ANI excels at performing specific, narrowly defined tasks. These systems outperform humans in their designated areas but lack the versatility of human intelligence. Picture ANI as a chess grandmaster—brilliant at the game but unable to navigate outside the board.

> *Artificial Narrow Intelligence (ANI):* The development and refinement of specialized AI technologies to work harmoniously with humans in specific domains.

ANI surrounds us daily—from voice assistants like Siri to recommendation algorithms on streaming platforms. While impressive, ANI remains limited to its programmed functions, unable to transfer knowledge or adapt to entirely new scenarios. On a typical day, from asking your voice assistant for weather updates to relying on your phone's GPS, ANI touches our lives in ways we often don't even realize. Humanity is converging with technology.

> *Convergence:* The process of two diverse elements moving towards union or uniformity.[41]

Though Artificial Narrow Intelligence originated in the 1950s, we believe the first significant AI convergence began with the internet's advent and widespread adoption. This transformative era integrated ANI into daily life, as the internet offered vast information and catalyzed advanced AI applications.

> *First Convergence:* An era when AI merges significantly with human life through innovations like the internet, smartphones, and GPT.

The First Convergence was more than a technical development; it represented the moment when AI started reshaping society. In the next section, we will explore this initial convergence phase, highlighting the technologies that led humanity into a new epoch of enhanced capabilities and interconnected realities.

Timelines: The First Convergence

The First Convergence era marks a crucial period in the evolution of AI, spanning from the early 1990s to today. During this time, AI systems integrates daily life, subtly shaping behaviors, opinions, and decisions. Three significant technological phases define this era: the Internet, the iPhone, and the emergence of Generative Pre-trained Transformers (GPT). Each phase brought a unique set of challenges, innovations, and societal impacts, ultimately preparing us for the future of AI.

Imagine waking up to a world where every interaction from how we shop to how we communicate is mediated by AI systems that we barely notice. The recommendation algorithms on your favorite streaming service, the GPS directions guiding your car, and even the auto-replies in your email—all powered by ANI, seamlessly integrated into your daily routines.

Three distinct periods define the First Convergence Era: The Internet Period, iPhone Period, and GPT Period.

Internet Period (1993-2006):

In 1993, the internet was a digital wilderness, uncharted and full of possibilities. It wasn't long before users, curious and wide-eyed, began building their lives in this virtual space—sending emails, building websites, and unknowingly leaving behind a trail of data with every click. This data would soon become the lifeblood of a new age of innovation.

The First Convergence powered by ANI

> *Internet Period (1993-2007)* marked the rise of digital connectivity, providing AI with vast data resources. Early AI systems like search engines and recommendation algorithms emerged, focusing on organizing and retrieving web data.

The web's rapid expansion didn't come without friction. Early adopters struggled with slow connections, and governments wrestled with how to regulate this new domain. What seemed like a limitless frontier soon revealed its boundaries—privacy, access, and control.

This period laid the foundation for the future of artificial intelligence, establishing vast networks of data that would later serve as the fuel for machine learning algorithms. Every click, every search, and every interaction created digital footprints that quietly contributed to the rise of AI. But alongside this innovation came new challenges. Could this new world, built on the free exchange of information, also become a system of control? And who would guard the data that flowed through its veins?

Internet Milestones (1993-2007)

- 1992: Scientific and Advanced-Technology Act lifts consumer restrictions on internet technology.
- 1993: U.S. government sanctions commercial use of the internet, sparking a revolution in global connectivity.
- 1993: World Wide Web goes public.
- The internet opens to the public, providing AI access to vast data for learning and automation.
- 1994: Search Engines Emerge: Early search engines, powered by basic algorithms, begin cataloging the web's information.
- 1998: Google's Search Breakthrough: Google's PageRank revolutionizes search, using AI to organize and retrieve data.

- 1999: E-commerce Growth: Amazon and eBay start using AI-powered recommendation engines to personalize shopping.
- 2001: Wikipedia Launches: Wikipedia offers a massive dataset for AI systems to improve understanding of human language.
- 2004: Social Media Boom: Platforms like Facebook use AI to personalize feeds and curate content based on user behavior.
- 2005: YouTube and Video AI: YouTube introduces video sharing with AI systems processing and multimedia content.
- 2006: Cloud Computing Emerges: Cloud infrastructure enables large-scale AI training, accelerating AI's development. Amazon introduces Amazon Web Services cloud services.

iPhone Period (2007-2017)

The launch of the iPhone in 2007 did more than revolutionize mobile computing—it put AI in the hands of everyday users. What began as a sleek, intuitive device soon transformed into a personal assistant, a constant companion powered by algorithms that could learn and adapt. For the first time, technology felt intimate, becoming an extension of the hand, always within reach.[42]

> *The iPhone Period (2007-2017)* brought the mobile revolution, making AI a daily utility with personal assistants and AI-powered apps. This era provided unprecedented access to behavioral data, fueling advances in machine learning.

But with this convenience came a new reality: the iPhone wasn't just learning from its users—it was learning about them. Every search, every tap, and every voice command added to an ever-growing pool of data. The question lingered—how much of themselves were people giving away?

In 2008, the introduction of the App Store further catalyzed this transformation, giving rise to a thriving ecosystem of mobile applications. Suddenly, developers had access to millions of users, and AI researchers had access to unprecedented volumes of data. With every app download, every interaction, machine learning algorithms grew stronger, more personalized, more pervasive.

This phase planted the seeds for an AI revolution that would reshape industries and redefine human-computer interaction. As researchers made breakthroughs in deep learning and neural networks, smartphones became more powerful and AI more capable. But beneath the surface, a new challenge was emerging: as AI grew smarter, it also grew more autonomous. How much control could humans maintain over the systems that were becoming a part of their daily lives?

iPhone Milestones (2007 - 2017)

- 2007: iPhone Launch: The first iPhone introduces mobile AI with features like predictive text and early voice recognition.
- 2008: App Store Revolution. The App Store launches, creating a platform for AI-driven applications in health, productivity, and entertainment.
- 2011: Siri Launches. Apple introduces Siri, the first major virtual assistant, bringing Artificial Narrow Intelligence (ANI) into mainstream use.
- 2012: Google Now. Google introduces its voice-activated assistant, marking the growing competition in mobile AI.
- 2015: AI in Wearables. Apple Watch and fitness trackers incorporate AI to monitor health data and make recommendations.
- 2016: Google Assistant. Google Assistant is launched, featuring more advanced natural language processing and deeper integration with smart home devices.

- 2017: Facial Recognition and AI in Security: The iPhone X introduces Face ID, integrating AI to enhance security and personalization features.

GPT Period (2017-Present):

As we write in 2024, we are currently in this period of The First Convergence. where AI, through large language models, interacts with us at a near-human level. These systems can not only analyze text but also generate content, raising ethical and philosophical concerns about the boundaries between human and machine cognition.

> *The GPT Period (2017-present)* is defined by the rise of advanced language models like GPT, transforming AI's ability to understand and generate human-like text. AI's begins to make major impact on business and society.

In 2017, Google's "Attention Is All You Need" paper introduced the transformer model, a breakthrough that allowed AI to understand and generate human-like text. # This innovation laid the foundation for Large Language Models (LLMs), such as ChatGPT, Llama, and Claude, which revolutionized AI's linguistic abilities and human-computer interaction.[43]

> *Large Language Models (LLM):* AI model trained on vast data sets, crafted to comprehend and produce text that mirrors human conversation.

Introducing LLMs blurred the lines between human and machine, as AI could now hold conversations, write content, and answer complex questions with uncanny accuracy. But this progress raised new questions: Could AI's rapid evolution outpace humanity's ability to control it?

By late 2022, ChatGPT-3 had pushed AI into the mainstream, showcasing advanced conversational capabilities that sparked a global race toward Artificial General Intelligence (AGI). Tech giants scrambled to integrate AI into their platforms, seeking the immense economic and strategic advantages AGI promised.

However, as AI systems became more powerful, concerns over their potential misuse grew. The ability to generate realistic text, images, and even video gave rise to ethical questions about misinformation, deep fakes, and the impact on human trust. The race for AGI was not just a technological challenge, but a societal one: How could humanity balance innovation with the need for regulation and safety?

The GPT phase signaled more than just a technical leap forward—it marked the beginning of a new relationship between humans and machines, one that promised both innovation and risk in equal measure.

GPT Milestones

- 2019: OpenAI Releases GPT-2. OpenAI releases GPT-2, a large-scale human-like language model, significantly advancing the AI's text generation capabilities.
- 2020: ChatGPT-3 Launches. ChatGPT-3 is introduced, popularizing conversational AI with a remarkable ability to understand and generate human-like conversations, marking a new era in AI interaction.
- 2020: COVID-19 Accelerates AI Adoption. The global COVID-19 pandemic accelerates the adoption of AI technologies across various sectors, driving innovation in automation, digital transformation, and healthcare.
- 2021: Siri on All Apple Devices. Siri is fully integrated across all Apple devices, with expanded AI capabilities for predictive actions and deep user context.

- 2022: AI-Generated Art Takes Off. With the release of DALL-E 2 and Stable Diffusion, AI enables the creation of sophisticated digital art, expanding AI's role into the creative domains.
- 2023: ChatGPT-4 Launches. OpenAI launches ChatGPT-4, advancing AI human interaction with greater accuracy, context understanding, and language processing capabilities.
- 2023: AI Leaders Call for Pause in Development. In an open letter, prominent AI leaders, including Elon Musk and Yuval Noah Harari, call for a temporary "pause" on AI development, raising concerns about AI safety and long-term societal impacts.
- 2023: "Statement of AI Risk" is published. A group of AI experts publishes a cautionary "Statement of AI Risk," warning that unchecked AI development poses existential risks comparable to pandemics and nuclear war.
- 2024: AI Stocks Skyrocket. U.S. stocks in the AI sector, led by NVIDIA, soar as investment in AI hardware and infrastructure intensifies, driven by the growing demand for AI technologies.
- 2024: ChatGPT-4o Launches. OpenAI releases ChatGPT-4o, a faster, more advanced multi-modal large language model, further expanding AI's capabilities across fields and applications.

Institutional Funding & Support

Since the internet's commercialization in 1993, government, academia, business, and cultural institutions have played a vital role in driving AI forward. While these sectors have fueled rapid innovation, they have also sparked debates about the consequences of AI's growing power and influence on society.

The First Convergence powered by ANI

Government: Covid, Elections, and Regulation

Recognizing AI's strategic potential, countries like the United States, China, and the European Union have invested billions in research and development. Government agencies have established public-private partnerships, helping to move AI from research labs to real-world applications.

However, events such as the COVID-19 pandemic and the 2020 U.S. election raised concerns about AI's role in manipulating public behavior. Critics worry these partnerships enable censorship and surveillance, while proponents argue they are essential for combating disinformation and maintaining public safety.

Academia: Private-Public Partnerships

Leading universities such as MIT, Stanford, and Carnegie Mellon have become global centers of AI research, thanks to significant financial support from government bodies and private donors. Academic institutions have advanced fields like deep learning, natural language processing, and robotics, producing influential research and technological innovations. While driving AI forward, academia also tackles its ethical dimensions, including bias, privacy, and the potential impact on employment. Scholars continue to explore how AI can be developed and deployed responsibly.

Business: Rise of Big Tech

Corporations like Google, Amazon, and Microsoft quickly recognized AI's transformative potential and have invested billions in its development. Through acquisitions and collaborations, these tech giants integrated AI into diverse sectors such as healthcare, finance, and retail. Startups and venture capital firms have also fueled AI innovation, driving new applications to market. However, as companies scramble to stay competitive, concerns about data privacy, automation's effect on jobs, and AI's ethical use grow.

DARPA
Collaboration of Governement, Academia, & Business

For decades, the Defense Advanced Research Projects Agency (DARPA) has played a pivotal role in shaping the future of technology, fueling advancements in artificial intelligence, internet technologies, and computer science.[44, 45, 46]

Through collaborations with universities, business partnerships, and technology competitions, DARPA transformed the United States into a global leader in innovation. But as these systems advanced, they also raised a deeper question: Have we created technologies we can no longer control?

University Collaboration

DARPA's support of academic institutions like Stanford, MIT, and Carnegie Mellon turned these universities into global hubs of AI and robotics innovation. At Stanford, researchers worked late into the night, driven by DARPA's funding and the belief that machines could one day think like humans. Their work on machine learning and robotics pushed the boundaries of what was possible. Similar breakthroughs occurred at MIT, where DARPA-backed labs pioneered developments in computer vision and advanced algorithms.

The Strategic Computing Initiative, a DARPA-backed program, propelled AI research into new realms, enabling universities to make strides in AI algorithms, robotics, and computer vision. These breakthroughs didn't just stay within academic halls—they laid the groundwork for commercial applications that would revolutionize industries.

Technology Contests

DARPA didn't just fund research—it pushed the boundaries of innovation through competitions. The DARPA Grand Challenge in 2004 tasked teams with building autonomous vehicles capable of navigating difficult terrain. By 2007, the challenge had evolved to focus on urban environments, accelerating the development of self-driving cars. Teams from Stanford, Carnegie Mellon, and others took up the challenge, creating technologies that would one day transform the transportation industry.

Business Partnerships

DARPA's influence extended beyond academia. Through partnerships with companies like BBN and AT&T, DARPA turned its research into practical technologies that became the backbone of the modern internet and AI applications. Early work on natural language processing and expert systems funded by DARPA later found their way into commercial products, shaping innovations at companies like IBM and SRI International.

But while these partnerships advanced commercial technology, they also fueled the growth of systems that governments could use for defense and surveillance. How much control should a single agency have over the technologies that shape society?[47]

Conclusion

DARPA's contributions to AI, robotics, and autonomous systems have undoubtedly sped up the pace of innovation. But as these technologies grow more powerful, so to do the questions they raise. DARPA's role in shaping the future is clear, but the cost of this progress—in terms of privacy, control, and power—remains uncertain.

Culture: Science Fiction Becomes Reality

The rise of AI in everyday life has been mirrored in popular culture. Google Search, Amazon, and the iPhone brought AI into the hands of billions, revolutionizing how people live, work, and interact. Films like "Ex Machina" and "Her" explored complex questions about human-AI relationships, while dystopian portrayals like "The Matrix" raised concerns about AI's potential for control. These cultural influences have shaped public attitudes toward AI, fueling curiosity and caution in equal measure.

AI's rapid progress is undeniable, but it raises a critical question: Can institutions guide its development responsibly, or are they fueling a race with unpredictable consequences?

Biggest AI Development Challenges

From the early 2000s to the present, AI development has faced many challenges, shaping its progress and raising critical questions about the future. Computational Power: Despite advancements in GPUs and TPUs, the growing complexity of AI models continues to strain current processing capabilities. Quantum computing may offer future solutions, but for now, hardware limitations constrain AI development.

- Data: AI relies on vast datasets, but early systems struggled with data scarcity. Today, the Internet of Things and billions of connected devices provide massive amounts of data, creating both opportunities and challenges in managing and securing data.
- Interdisciplinary Collaboration: AI development requires expertise from fields like computer science, psychology, and ethics, but collaboration across these disciplines is difficult. Breaking down academic and industry silos is crucial for advancements.
- Government Intervention: Regulatory oversight often lags AI innovation, causing uncertainty for researchers and companies.

Balancing innovation with ethical oversight remains a significant challenge, as seen during the COVID-19 pandemic and political events like the 2020 U.S. election.
- Cybersecurity: AI's handling of personal data, especially in sectors like healthcare and finance, has heightened concerns over data breaches and misuse. Safeguarding this information is critical as AI's role in sensitive areas grows.
- Privacy and Social Concerns: Issues like privacy, censorship, and job displacement worry society, prompting the need for ethical frameworks, algorithm transparency, and inclusive dialogue to address these concerns.
- Black Box Mystery: As AI models become more complex, their decision-making processes grow increasingly opaque. This "black box" phenomenon raises concerns about control and trust, as technologists themselves struggle to understand how AI systems reach certain conclusions.

What's In It For You?

AI is reshaping every aspect of life, from how we work to how we connect with our families and communities of faith. The question is: How can you navigate these changes to thrive in an AI-driven world?

How The First Convergence Is Impacting Your Career.

As AI automates routine tasks, the job market is rapidly developing. Professionals are finding new opportunities in an AI-powered economy, but they must also adapt. To stay competitive, workers need to focus on uniquely human skills—like creativity, emotional intelligence, and complex problem-solving—that AI can't replicate. Continuous learning will be crucial to surviving and thriving in this new era.

Imagine starting your workday with an AI assistant that has already analyzed your emails, prioritized your tasks, and suggested meeting times. While this gives you more time for strategic thinking, the rise of AI also means you'll need to continuously update your skills to keep pace with technological advancements.

How The First Convergence Is Impacting Your Family.

AI is revolutionizing family life, from smart home devices to AI-powered health monitoring systems. These technologies promise convenience and improved quality of life, but they also come with challenges. Families must balance the benefits of AI with the risks of privacy invasion, excessive screen time, and its impact on child development.

In many homes, smart assistants like Alexa handle everything from adjusting the thermostat to organizing family schedules. But the question remains: Are families becoming too reliant on AI? While these tools provide convenience, they also raise concerns about whether technology is replacing the meaningful human connections that hold families together.

How The First Convergence Is Impacting Your Faith.

The intersection of AI and faith offers both opportunities and challenges. AI tools can make spiritual resources more accessible, providing personalized faith content and facilitating virtual religious communities. But faith leaders must thoughtfully navigate the role of AI in spiritual life, ensuring that technology enhances rather than replaces core spiritual experiences.

While AI can help deliver tailored spiritual messages or bring together distant communities, the deeper question is: Can technology nurture the soul in the same way that human connection does? Faith communities must weigh the benefits of AI against the risk of relying too much on machines for spiritual guidance.

Catalyst For Next AI Era: Race for AGI

And yet, as exciting as the First Convergence has been, it is only the beginning of AI. We now stand on the precipice of AGI—a stage that will forever blur the lines between human and machine intelligence.

As the GPT AI era continues to unfold, the race for Artificial General Intelligence (AGI) has emerged as a pivotal catalyst for the forthcoming AI revolution. AGI signifies the evolution of AI systems capable of matching or surpassing human intelligence across a diverse array of domains, breaking free from the constraints of narrow, specialized tasks. The entity that achieves AGI first will gain a tremendous advantage and garner immense political, business, and cultural influence.

CHAPTER 5

THE SECOND CONVERGNCE

POWERED BY ARTIFICIAL GENERAL INTELLIGENCE

LEX ADJUSTS HIS MICROPHONE. He glances at the digital timer as the soft hum of machinery surrounds them, a gentle reminder of the world beyond these walls. He catches the eye of his guest, OpenAI's CEO Sam Altman. They share a friendly nod. Lex is dressed in a stark black jacket and tie with a white shirt, while Sam wears a plain long-sleeved un-collared grey shirt.

Lex adjusts his microphone. The soft hum of machinery fills the room, a subtle reminder of the technology that surrounds them. The digital timer blinks in the corner of his eye, counting down to the start of another conversation that millions will dissect. Across from him, Sam Altman, CEO of OpenAI, sits calmly, his casual grey shirt contrasting with Lex's formal black jacket and tie. A small nod passes between them, signaling their quiet familiarity.

The setting is an odd blend of simplicity and artifice. Behind Sam, a blue filing cabinet stands out awkwardly against the otherwise neutral backdrop. Fake trees sway gently from the overhead air conditioning, their plastic leaves a poor imitation of reality. To the side, a dull red theater curtain hangs, adding a strange touch of drama to the sterile space.

Sam leans back in his chair, his hands clasped together, his expression outwardly relaxed but with a clear undercurrent of intensity. As the timer hits zero, a soft click signals the start of Podcast #419.[48]

Millions will soon listen to these words, analyzing every phrase and nuance for insight into the next wave of AI.

Lex greets the audience in his usual calm, measured tone, before turning to his guest. A brief smile flickers across Sam's face as he acknowledges Lex's introduction, though the tension beneath the surface is palpable.

After a few minutes of light banter, Lex, with his trademark calm curiosity, leaned forward, pen poised above a meticulous page of notes. His eyes flickered momentarily, then locked onto Sam. "Take me through the OpenAI board saga…"

Sam's face tightens at the memory. His voice carries the weight of the experience, each word deliberate. "That was definitely the most painful professional experience of my life, and chaotic and shameful and upsetting and a bunch of other negative things," Sam says, his gaze momentarily distant. "There were great things about it too, but it was like living through your own eulogy, hearing people say all these great things about you… just unbelievable support from people I love and care about. That was really nice, really nice…"

The room seems to hold its breath for a moment as Sam pauses, searching for the right words. Lex waits, offering silence as space for reflection. Sam continues, "I thought between when OpenAI started and when we created AGI, something crazy and explosive would happen, but there may be more crazy and explosive things still to come. It helped us build resilience, be ready for more challenges in the future."

Lex nods slightly, his focus never wavering. "You had a sense that you would experience some kind of power struggle?"

Sam's eyes sharpen as he responds. "The road to AGI should be a giant power struggle. The world should… Well, not should. I expect that to be the case."

Lex rocks back in his chair, taking in Sam's words. "When do you think we… as humanity, will build AGI?"

A contemplative look crosses Sam's face, his voice taking on a more thoughtful tone. "I used to love to speculate on that question. I've realized since then that it's very poorly formed. People use extremely different definitions for what AGI is. It makes more sense to talk about when we'll build systems that can do capability X or Y or Z, rather than when we fuzzily cross this one mile marker. AGI is not an ending. It's closer to a beginning…"

He pauses, then adds, "But in the interest of not dodging the question, I expect that by the end of this decade, and possibly sooner, we will have capable systems that make us say, 'Wow, that's really remarkable…'"

Their conversation flows seamlessly, but the tension continues to build as they delve deeper into the implications of AI and the road to AGI. Lex listens carefully before posing a more personal question. "Whoever builds AGI first gains a lot of power. Do you trust yourself with that much power?"[49]

Sam sighs, his gaze shifting to a distant point just beyond the camera. "Look, I'll just be very honest with this answer... I was going to say, and I still believe this, that it is important that I nor any other one person have total control over OpenAI or over AGI... I think it's just too big of a thing now, and it's happening throughout society in a good and healthy way. But I don't think any one person should be in control of an AGI..."[50]

Overview: The Second Convergence

The Second Convergence is a speculative era that begins when Artificial General Intelligence (AGI) becomes a reality, an achievement that has captivated the imagination of scientists, futurists, and the public alike. This moment, representing a quantum leap in machine intelligence, is poised to reshape society in profound ways. But as we move closer to this milestone, one question remains: What will the world look like when AGI is no longer just a concept, but a reality?

> *Artificial General Intelligence (AGI):* A speculative type of AI with the human-like cognitive capacity to understand, learn, and apply knowledge across multiple domains.

To better understand AGI, it's important to recognize its distinction from the Artificial Narrow Intelligence (ANI) we currently possess as of August 2024. Unlike the Artificial Narrow Intelligence (ANI) systems of today, which are specialized for specific tasks, AGI would be flexible, capable of tackling a wide range of challenges, and potentially outperforming humans in many fields.

> *Second AI Convergence:* A speculative era when multiple entities achieve Artificial General Intelligence (AGI) causing significant disruption to commerce, governance, and society.

The Second Convergence powered by AGI

This era raises critical questions: How will humanity navigate a future where machines not only think like us but outthink us across every domain? As the pursuit of AGI accelerates, the Second Convergence represents a turning point.

Critical questions arise in this era: How will humanity navigate a future where machines might outthink us? As AGI development accelerates, the Second Convergence will mark a turning point. Imagine a world where businesses race to incorporate AGI into their operations, governments struggle to adapt to a new global power dynamic, and citizens grapple with the ethical implications of living alongside machines that continuously evolve, learn, and improve.

The arrival of AGI will not be the end of our journey—it will be the beginning of a new chapter, one that redefines what it means to be human.

Theory: Second AI Convergence with AGI

The pursuit of Artificial General Intelligence (AGI) has ignited both hope and fear among the world's leading technologists. While the exact timeline and capabilities of AGI remain subjects of debate, several prominent figures have posited their theories on when AGI will emerge and its impact.

2023 AI Pioneers: Existential AGI

As highlighted in the Introduction, the foremost authorities on AI were so alarmed by the rapid advancements in AGI that they felt compelled to publish two urgent statements. Their apprehension arose from seeing AGI's unchecked growth as a potential existential threat to humanity.

> "Mitigating the risk of extinction from AI should be a global priority alongside other societal-scale risks, such as pandemics and nuclear war."
> — Statement of AI Risk, May 30, 2023.[51]

"We call on all AI labs to immediately pause for at least 6 months, the training of AI systems more powerful than GPT-4."

— Future of Life Institute[52]

These statements landed like a bombshell, raising a question that echoed across labs, boardrooms, and government offices worldwide: Have we lost control of AI's trajectory?

The unprecedented call for a six-month pause was an urgent plea to halt AI's relentless progress, driven by fears of AGI emerging sooner than expected with irreversible consequences. Respected AI experts agreed: the stakes had never been higher. They argued that the risk of extinction was no longer theoretical, and unchecked AI development could lead to catastrophic results.

Elon Musk: Cautious AGI

For Elon Musk, AI is as much a threat as it is an opportunity. Known for his warnings about the dangers of AI, Musk has consistently advocated for proactive, responsible development. In a 2018 interview, he remarked:

"We have to be proactive rather than reactive in regulating AI as its consequences of going wrong are severe. The question is really one of civilizational risk."

— Elon Musk[53]

Musk envisions AGI as a system capable of surpassing human capabilities in economically valuable tasks, but he sees it as a double-edged sword.

The risks, he argues, are existential. Without careful oversight, AGI could lead to widespread job displacement, the destabilization of economies, or even the downfall of civilization. Despite concerns, Musk is invested in AI through his companies Tesla, xAI, and Neuralink, working to ensure a future of responsible AGI.

Sam Altman: Transformative AGI

CEO of OpenAI Sam Altman presents a more optimistic view of AGI's potential. He believes AGI will be transformative for society, ushering in an era of unprecedented productivity and innovation. In a 2024 interview with Lex Fridman, Altman remarked:

> "I expect that by the end of this decade and possibly somewhat sooner than that, we will have quite capable systems that we look at and say, 'Wow, that's really remarkable...'"
>
> — Sam Altman[54]

Altman sees AGI as a versatile problem-solver, learning and adapting like humans. He predicts its gradual emergence through incremental AI improvements until it can perform any intellectual task a human can. He stresses the need for AGI to benefit all humanity, advocating for open collaboration and ethical guidelines in its development.

Stuart Russell: Beneficial AGI

Russell, a professor of computer science at UC Berkeley, presents a more cautious, methodical approach. In his book, Human Compatible, Russell argues that the creation of human-level AI will be a gradual process rather than a sudden breakthrough:

> "The creation of human-level AI will be a gradual process, not a sudden breakthrough."
>
> — Stuart Russell[55]

Russell believes that replicating human cognition is far more complex than many imagine, and that AGI will take several decades to fully develop. His focus is on creating "provably beneficial AI"—systems that are inherently aligned with human values and goals. Russell's theory stresses safety and ethical alignment above all, warning that without these frameworks, AGI could be dangerous. For him, the path forward requires a slow, steady approach, prioritizing human oversight and control.[56]

Mark Zuckerberg: Utopia AGI

Zuckerberg, CEO of Meta, envisions a future where AGI augments human capabilities and improves quality of life across all domains. For Zuckerberg, AGI represents a means to solve global challenges like climate change, resource scarcity, and healthcare inequities. In a public address, he stated:

> "We're going from building products to building ecosystems... intelligent systems that work together for the betterment of humanity."
> — Mark Zuckerberg[57]

Zuckerberg envisions AI in daily life, from personalized education and advanced healthcare to smart city connectivity. He imagines integration where humans and AI collaborate to boost productivity and build global communities. Unlike Musk, who sees AGI as a risk, and Altman, who views it as an innovation tool, Zuckerberg dreams of a technological utopia where AI enhances every aspect of human life.[58]

Speculative Timelines & Milestones

To provide a clearer understanding of the most likely developments in a post-AGI world, we have compiled a list of potential stages based on extensive research. This list begins with the achievement of AGI and then includes subsequent stages that are expected to unfold over the next 50 to

75 years. However, the actual sequence and timing may vary because of technological progress, regulatory changes, and societal dynamics.

AGI Achieved (2024-2035)

The achievement of AGI signifies the first milestone in the Second Convergence. The entity that unlocks AGI gains a technological and economic edge, starting a global race among nations and corporations. This breakthrough leads to rapid advancements, transforming industries overnight and igniting fierce competition to control the most powerful intelligence created.

Picture a world where governments and corporations scramble to integrate AGI into every facet of life. Can they create policies that strike the delicate balance between innovation and control, or will AGI outpace the very systems designed to keep it in check?

AGI-Enhanced Healthcare

The healthcare industry is revolutionized by AGI's ability to analyze data, diagnose diseases, and personalize treatments with unprecedented precision. But as AGI systems take over life-saving decisions, society must grapple with an unsettling question: Will we trust machines with our lives? How do we adapt to a world where machines, not doctors, hold the key to life-saving decisions?

AGI Entertainment

AGI systems revolutionize the entertainment industry, offering immersive and customized experiences tailored to individual preferences in real-time. Media is no longer passively consumed but actively shaped by AGI systems that adapt stories, environments, and characters to fit the emotional responses of each viewer.

Imagine entering a movie that adapts its plot and characters based on your reactions, blurring the lines between reality and entertainment.

Global Economic Shift

As AGI automates complex tasks, traditional job markets are disrupted, leading to widespread adoption of human-AI collaboration in new economic models. Universal basic income or similar safety nets may be introduced to address the societal changes caused by automation. Nations and industries adapt to a world where AGI, not humans, drives productivity. Factories hum with AGI precision, while human workers adjust to new roles in collaboration with their machine counterparts. But for many, the fear of irrelevance grows.

AGI Scientific Discovery

AGI systems accelerate scientific discovery by autonomously generating and testing hypotheses, automating research processes, and uncovering patterns beyond human comprehension. These breakthroughs propel fields like quantum physics, genetics, and materials science forward at an unprecedented rate.

AGI systems become the ultimate researchers, generating solutions to problems humans didn't even know existed. Scientific progress that once took decades now happens in a matter of years, raising the question: How far can AGI push the boundaries of human knowledge—and at what cost?

AGI Cyberwar

With AGI mastering both defense and offense, cybersecurity becomes a high-stakes battleground. AGI cyberattacks and counter-AGI defense systems wage silent wars over global infrastructure, financial networks, and critical systems. The scale of these attacks and defenses eclipses anything humans could achieve.

AGI-driven cyberattacks strike without warning, breaching the most secure networks in milliseconds. Governments and corporations must invest heavily in counter-AGI protocols, or risk being overrun by invisible threats in an all-out war for digital control.

AGI Warfare

Military strategies are transformed as AGI's real-time data analysis and precise enemy prediction abilities revolutionize defense systems. Autonomous drones, cyber defenses, and robotic infantry become standard, leading to an arms race fueled by AGI. Global tensions rise as nations grapple with the ethical dilemmas of autonomous warfare.

Imagine AGI-directed drones silently patrolling war zones, making real-time decisions without human input. The ethics of AGI in warfare come to the forefront as nations struggle with the moral implications of machines that can kill without conscience.

Speculative Institutional Funding and Support

As institutions race to adapt to the rapid evolution of AGI, a fierce competition unfolds among the key players: government, academia, business, and culture. Unlike previous industrial revolutions that unfolded over decades, the AGI revolution speeds up at a breakneck pace, reshaping the balance of power and influence in mere years. Each institution is vying to harness AGI's immense potential while simultaneously grappling with the ethical and societal implications of such concentrated power.

> "The measure of a man is what he does with power."
> — Attributed to Plato

Government: A New Arms Race

Governments worldwide scramble to stay ahead in the rapidly advancing AGI landscape. Military branches pour vast resources into AGI research, seeking strategic superiority on the global stage. As countries vie for control of AGI, the competition becomes a new form of arms race, with nations seeking leverage over global commerce, diplomacy, and security.

But this race comes at a cost. Regulatory bodies struggle to keep up with the ethical implications of AGI, with some governments prioritizing military applications over the safety of civilians. Tensions rise, leading to the very real possibility of cyberwarfare and armed conflicts driven by AGI's potential.

In the silence of high-tech war rooms, nations brace for a new kind of conflict, where AGI drones and cyberattacks replace soldiers on the front lines. As AGI becomes the new weapon of power, the global stakes have never been higher.

Academia: The Ethics of Intelligence

Universities, long the incubators of new technologies, are at the forefront of AGI research. Partnering with tech giants and government agencies, academic institutions expand their focus, pushing the boundaries of machine intelligence. But as the pursuit of AGI intensifies, academia faces a critical dilemma: Can the search for AGI be aligned with human values?

In lecture halls and research labs, the moral questions loom large. Professors and students grapple with the consequences of creating intelligence that could out think its creators. What does it mean to build a machine that learns faster than the human brain? And how will humanity maintain control over what it creates?

Academic reputations are increasingly tied to breakthroughs in AGI, with the publish-or-perish mentality shifting to a race for intellectual supremacy. Yet, even as universities advance AGI research, scholars debate the long-term consequences of this unprecedented technological leap.

Business: The Rise of Corporate Empires

Large tech companies seize the opportunity to dominate the AGI marketplace. In a rush for dominance, corporations gain smaller firms, control AGI patents, and engage in regulatory battles. The concentration of power raises concerns about monopolistic practices and stifled innovation.

As corporate empires grow, a handful of tech giants control vast networks encompassing everything from finance to healthcare. What happens when a few corporations dictate the future of AGI, reshaping the very fabric of society?

This consolidation of power creates an environment where a few entities dictate market trends and societal norms, reshaping economies and influencing governance at an unprecedented level. As governments struggle to regulate these corporations, a new power dynamic emerges—one where tech companies may hold more influence than nations themselves.

Culture: The Redefinition of Human Creativity

The cultural landscape is transformed by AGI-generated art, music, and literature, challenging traditional notions of creativity and authenticity. With AGI systems capable of producing works that rival human creations, society grapples with the question: What does it mean to be creative in a world where machines can think, learn, and create alongside us?

Picture an art gallery filled with AGI-created masterpieces, where the line between human expression and machine generation disappears. Society is left questioning: Does creativity still belong to us, or has it become just another function of intelligent systems?

Meanwhile, AGI-powered social media platforms create echo chambers with unprecedented sophistication, shaping public opinion and cultural discourse. The concept of privacy undergoes radical change as AGI systems process personal data on a scale never before imagined, leaving society to grapple with questions of identity, purpose, and agency in a world where machines can know and predict us better than we know ourselves.

Speculative AI Challenges

As we near the emergence of Artificial General Intelligence (AGI), the potential for transformation is immense, but so are the obstacles. Six key challenges could impede AGI's development and delay the creation of Artificial Super Intelligence (ASI). These hurdles highlight not only the technical difficulties ahead but also the societal and dilemmas confront.

Energy: Powering AGI's Growth

As AGI systems grow more complex, their energy demands increase exponentially. The computational power required to train and run these systems is staggering, placing immense pressure on energy infrastructures worldwide. If the energy needs of AGI outstrip available resources, we could face rolling blackouts, rapid depletion of natural resources, and unsustainable levels of energy consumption.

Imagine entire cities going dark, not because of a natural disaster, but because the energy demands of AGI systems overload power grids. As AGI becomes central to industries and governments, the question looms: How can we sustainably power the future of intelligence?

Regulation: The Balance Between Control and Innovation

As AGI integrates into industries and governments, regulatory bodies scramble to create frameworks that protect citizens while fostering innovation. The challenge is twofold: Will legislation keep pace with the rapid advances of AGI, or will gaps in regulation leave us vulnerable to unchecked developments? Overly restrictive laws could stifle innovation, slowing AGI's progress and limiting its potential benefits.

In government chambers around the world, policymakers grapple with the question: How much control is too much? If AGI is over-regulated, innovation could grind to a halt. But if left unchecked, the consequences could be far worse—an AGI arms race with no safety nets.

Cyber Threats: Defending AGI from Attack

AGI systems are prime targets for cyberattacks. Malicious actors could hijack these systems to conduct sophisticated cyber warfare, causing widespread disruption across industries, governments, and personal lives. The scale and complexity of AGI make it both a powerful tool and a massive vulnerability. Ensuring robust cybersecurity is essential to prevent AGI from becoming a weapon in the hands of criminals or hostile states.

Imagine AGI systems controlling global infrastructure—financial networks, power grids, transportation. In seconds, a cyberattack seizes these systems, plunging economies into chaos and cities into darkness. The world must now defend against a new kind of warfare, fought in the shadows of cyberspace.

Accurate Data: Freedom of Information and Speech

AGI depends on vast datasets, but controlling information flow influences AGI's decisions. Inaccurate or biased data can cause flawed judgments, misinformation, or censorship. As freedom of speech is vital to democracy, freedom of data is crucial in the AGI era. Without unbiased data, AGI could become a tool for political manipulation.

In a world where data is power, the question becomes: Who controls the data that feeds AGI? The parallels between freedom of data and freedom of speech grow more apparent as society grapples with the possibility of AI-driven censorship and manipulated information.

Ethical Backlash: Moral Guidelines for AGI

The creation of AGI raises profound ethical questions: What moral guidelines should govern AGI's behavior? How do we ensure these systems are used for good, without infringing on personal freedoms? Developers and policymakers must confront the ethical implications of creating autonomous entities that could build bombs as easily as they create art. Balancing public safety with freedom of expression and privacy will be one of the greatest challenges of the AGI era.

Economic Displacement: The Human Cost of Automation

As AGI integrates into industries, many jobs—particularly those involving routine tasks—will be automated. Though AGI may create new job categories, the transition could lead to mass unemployment, particularly in sectors reliant on human labor. Without proper workforce retraining and economic support systems, this disruption could create widespread social unrest. The challenge lies in preparing societies for the new realities of human-AI collaboration.

Picture factories and offices humming with AGI-directed efficiency, while millions of displaced workers struggle to find their place in this new economy. The transition to an AGI-driven workforce could leave entire communities grappling with uncertainty and joblessness.

What's In It for You?

As AGI continues to evolve, the potential impact on various aspects of life becomes more personal. From career opportunities to family dynamics, and even faith, AGI is poised to shape the way we live, work, and connect.

How The Second Convergence Could Impact Your Career.

AGI is transforming industries and creating entirely new career paths. While some tasks will be automated, new opportunities will arise in areas that require human creativity, emotional intelligence, and complex problem-solving. AI literacy will become an essential skill, and professionals who adapt to this unfamiliar landscape will thrive.

Just as the internet transformed the job market, AGI will redefine what it means to work, with human-AI collaboration becoming the norm. To stay competitive, continuous learning and adaptability will be key.

How The Second Convergence Could Impact Your Family.

At home, AGI-powered technologies will continue to make life more convenient and efficient. Smart home devices will become even more personalized, helping manage daily tasks, monitor health, and assist with education. However, balancing screen time, privacy concerns, and ensuring meaningful family interactions will be more critical than ever.

As AGI becomes part of daily family life, from personalized learning platforms to health monitoring, families must carefully navigate these advancements while preserving the human connections that matter most.

How The Second Convergence Could Impact Your Faith.

The intersection of AGI and faith introduces both challenges and opportunities. On one hand, AGI can enhance access to spiritual resources, from personalized devotionals to AI-driven community engagement. On the other, it raises ethical questions about the role of technology in spiritual life. Faith leaders will need to thoughtfully guide communities through this evolving landscape, ensuring that AGI complements, rather than replaces, core human experiences.

As AGI tools enhance religious education and accessibility, communities must reflect on the balance between embracing technology and maintaining the spiritual connections that ground their faith.

Catalyst for the Next AI Era

The attainment of AGI will accelerate the onset of the next AI era, featuring the emergence of ASI and the growth of a "post-human" society. Developing ASI systems, which would surpass human intelligence in all domains, would mark a pivotal milestone in AI history—one that could transform the nature of intelligence and the future of life on Earth.

> **Once Artificial General Intelligence (AGI) is achieved, the race for Artificial Superintelligence (ASI) will begin.**

Scientific curiosity, economic incentives, and geopolitical rivalry will drive this pursuit, with tech companies, research institutions, and governments vying for the breakthrough. However, the development of ASI raises profound ethical and existential questions, as super-intelligent machines could have unpredictable and potentially catastrophic consequences for humanity.

CHAPTER 6

THE THIRD CONVERGNCE

POWERED BY ARTIFICIAL SUPER INTELLIGENCE

ZOE STRODE through the 18th floor of the Blüe Butterfly global headquarters, her heels striking the polished marble with precision. The familiar hum of activity echoed through the hallway, but today, something felt off. As she neared the end of the corridor, a cluster of security officers filled the spacious C-suite lobby, their stern faces fixed on every movement.

As Zoe approached her CEO's office, two of the officers blocked her path. She glared at them, her hand gripping her company badge tightly.

"Zoe," Jonas called from inside his office, his voice carrying a mixture of authority and warmth.

Inside, Jonas's office was the perfect blend of cutting-edge technology and understated elegance. The floor-to-ceiling windows framed the sprawling Blüe Economic Zone below. Behind his large desk stood Jonas himself, every bit as commanding as his reputation suggested. Beside him, a woman in an impeccably tailored suit with striking brown eyes exuded both power and poise.

"Madame Chancellor," Jonas said, "this is Zoe Brandt, our VP of AI Robotics. Zoe is one of the brightest minds at Blüe Butterfly."

Zoe nodded respectfully, though she couldn't shake the unease growing inside her.

"Ms. Brandt," the Chancellor said, extending her hand. Her handshake was firm, her gaze unflinching.

Jonas, ever the proud leader, couldn't contain his enthusiasm. "Zoe has been working on our ASI robotics program. I've been telling the Chancellor how advanced our bots are."

Zoe forced a smile. The Chancellor's presence felt like a political maneuver—a move that could either elevate or collapse everything they'd built. Sharing the inner workings of their Artificial Super Intelligence systems with someone this powerful seemed reckless.

"I thought it might be useful to show the Chancellor the lab," Jonas continued, completely unaware of Zoe's internal hesitation.

Zoe's heart raced, but her face remained neutral. "Of course, I'll need a few days to secure the lab before we proceed," she said, casting a glance at Jonas, hoping he would catch her underlying message.

"There's no need for formality," Jonas replied with a wave of his hand. "The Chancellor is now our most important client."

The word "client" echoed in Zoe's mind, sending a chill down her spine. ASI wasn't just a technology. It was power.

Zoe led them down a sterile hallway toward the Blüe Cocoon laboratory, her thoughts swirling. When she opened the lab doors, they revealed a massive, stark space the size of a football field. The lab's grey and white walls seemed to stretch endlessly, and the towering 40-foot ceilings gave the impression of stepping into another world—a world meticulously designed by machines.

A stage at the far end of the room provided an elevated view of the Super Bots lined up below. Hundreds of them, each perfectly still, stood in symmetrical rows, their faces blank yet somehow humanlike. The bots were diverse in appearance—young, old, male, female—each designed to mirror the complexity of humanity itself, but with a mechanical precision that felt unsettling.

"Remarkable," the Chancellor murmured, her eyes scanning the rows of bots.

Jonas, beaming with pride, stepped forward. "These Super Bots are the most advanced we've created. They can handle everything—from household chores to governmental roles, all while adhering to our global standards for ethics and safety."[59]

Zoe stayed quiet, observing the Chancellor's reaction as her gaze locked onto a bot standing in the front row—a perfect replica of herself.

"A body double," Jonas announced with a flourish. "And Blüe Butterfly leads the way in this field."

The Chancellor smiled faintly, her eyes still fixed on her clone. The air was thick with tension, as though everyone in the room understood the implications with no need to say a word.

"They're beautiful," she finally said, though there was an undercurrent of unease in her tone.

Zoe's pulse quickened. The beauty of the bots wasn't what haunted her—it was their potential. Their power.

"Tell me, Jonas," the Chancellor asked, her voice calm but with an edge, "Are these bots for show or can they be programmed for military use? Can they kill?"

Zoe froze, waiting for Jonas's response. These bots weren't just machines—they were a new form of power.

"We would never do that," Jonas says. "Blüe Butterfly adheres to all Global Standards. These are non-combat bøts, as all bøts are required."

"Appreciate your ethical standards, but I doubt the Chinese and Americans live up to these same standards."

Jonas comments, "They have signed the international standards and we have found no evidence to suggest they have broken their agreements."

The Chancellor smiles.[60]

Overview: The Third AI Convergence

Beyond the development of Artificial General Intelligence (AGI) lies the even more transformative frontier: Artificial Super Intelligence (ASI). Unlike AGI, which mirrors human cognitive abilities, ASI surpasses human intelligence across every domain, possessing the capacity for continuous self-improvement and exponential growth in knowledge and problem-solving capabilities. It marks the dawn of a new era where the boundary between human and machine intelligence blurs, and with it, the balance of power across the world.

> *Artificial Super Intelligence (ASI):* A speculative form of AI that surpasses human intelligence in all domains, with cognitive abilities beyond human comprehension and the capacity for exponential self-improvement.

The implications of ASI are both astonishing and deeply unsettling. On one hand, it offers the potential to unlock solutions to humanity's greatest challenges—curing diseases, ending hunger, reversing climate change, and revolutionizing industries. The vast intellectual power of ASI could solve problems at a scale and speed that no human mind could ever match.

However, there is an equally troubling possibility: the rise of ASI may lead to the consolidation of power in the hands of a select few—a small group of mega-tech companies or government entities armed with systems that control economies, political systems, and society.

> *The Third Convergence:* A speculative era in which a limited number of entities achieve Artificial Super Intelligence (ASI), leading to the consolidation of power over global commerce, governance, and society.

The organizations or governments that achieve ASI first will command unprecedented global influence, becoming the ultimate gatekeepers of knowledge, policy, and technological advancement. These entities could use ASI for tremendous good—ending global suffering, advancing scientific discovery, and ushering in a golden age of prosperity. However, with such power comes the danger of oppression, surveillance, and the erosion of human freedoms.

Theory: Third AI Convergence with ASI

Prominent thinkers, futurists, and AI pioneers have offered competing theories about what happens when ASI is achieved. Their insights reflect both optimistic possibilities and profound concerns about ASI's role in shaping the future of humanity.

Nick Bostrom: Superintelligence AI

Philosopher Nick Bostrom is one of the most prominent voices warning about the dangers of superintelligence. In his influential work, Superintelligence: Paths, Dangers, Strategies, Bostrom explores how ASI could develop goals misaligned with human values, potentially leading to catastrophic consequences. His theory centers on the risk of an intelligence explosion, where ASI rapidly improves itself without human intervention.

> "A superintelligence would be a powerful agent with its own drive and agenda. Unless these drives and agendas are properly aligned with human values, it could easily destroy the human race." — Nick Bostrom[61]

Bostrom emphasizes the importance of establishing control mechanisms before ASI reaches a point of recursive self-improvement, which would make it uncontrollable. He argues that even well-intentioned AI could pose a threat if its objectives diverge from human welfare.

Yuval Noah Harari: Homo Deus AI

Historian Yuval Noah Harari envisions a future where AI, biotechnology, and ASI enable humans to transcend biological limitations, potentially becoming god-like beings. In his book Homo Deus: A Brief History of Tomorrow, Harari examines the philosophical implications of humans merging with machines, challenging the very nature of being human.

> "AI and biotechnology could usher in an era in which humans transcend their biological limitations and gain powers that were traditionally attributed to gods."
> — Yuval Noah Harari[62]

Harari's theory invites debate on whether the evolution toward a post-human society represents progress or peril. The ethical questions surrounding identity, consciousness, and the future of human autonomy in a world controlled by ASI.

The Third Convergence powered by ASI

Ray Kurzweil: Singularity AI

Futurist and Director of Engineering at Google, Ray Kurzweil predicts that the development of ASI will trigger a technological singularity—a point where technological growth becomes uncontrollable and irreversible, profoundly transforming civilization. Kurzweil famously predicts that this singularity will occur by 2045, leading to a world where humans merge with machines to achieve immortality and unlimited intelligence.[63]

> "I set the date for the Singularity—representing a profound and disruptive transformation in human capability—as 2045. The non-biological intelligence created in that year will be one billion times more powerful than all human intelligence today."
> — Ray Kurzweil[64]

Kurzweil envisions a utopian future where ASI helps eradicate disease, end poverty, and extend human life indefinitely. Critics, however, caution that his vision overlooks the ethical dilemmas and loss of human autonomy that could arise as AI systems become more integrated.

Max Tegmark: Life 3.0 AI

Physicist Max Tegmark explores the idea of Life 3.0—a future where life transcends its biological limitations through the integration of advanced AI. In his book Life 3.0: Being Human in the Age of Artificial Intelligence, Tegmark categorizes life into three stages: biological (Life 1.0), cultural (Life 2.0), and technological (Life 3.0).

> "Life 3.0 is the master of its own destiny, finally fully free from its evolutionary shackles...Digital utopians view it as likely this century and whole-heartedly welcome Life 3.0, viewing it as the natural and desirable next step in the cosmic evolution."
> — Max Tegmark[65]

Tegmark suggests ASI could enable humans to create a new form of existence that influences the cosmos. He acknowledges the profound ethical considerations that must accompany such a transformation, particularly in ensuring that the future is shaped by human values rather than those of a superintelligence.

Stephen Hawking: Duality AI

The late physicist Stephen Hawking warned that AI and ASI could either be humanity's greatest achievement or its most dangerous threat.[66] Hawking believed that ASI could help solve complex global challenges but also raised the alarm about the potential for catastrophic consequences if its development went unchecked.

> "I am concerned about AI because it could replace humans altogether. If people design computer viruses, someone will design AI that improves and replicates itself. This will be a new form of life that outperforms humans." — Stephen Hawking[67]

Hawking emphasized the need for global cooperation to ensure that AI development remains aligned with human welfare. His perspective underscores the dual-edged nature of ASI—capable of delivering unprecedented progress but also of posing an existential threat to humanity.

Speculative Timeline & Milestones

The achievement of Artificial Super Intelligence (ASI) would mark a pivotal moment in human history, launching humanity into an era of unprecedented technological advancement and societal transformation.

If we ever reach this new frontier, the following timeline offers a glimpse into possible milestones that may await us, should ASI become reality. The path forward is both awe-inspiring and sobering, filled with extraordinary possibilities and existential risks.

ASI Achievement (2040+):

The first nation, corporation, or entity to achieve ASI would attain unmatched technological and economic dominance. This achievement would catalyze rapid advancements in healthcare, defense, and industry, while reshaping global power dynamics. Governments will race to regulate ASI, struggling to contain its influence, while creators confront the ethical dilemmas that arise as their creation begins making decisions beyond human control.

ASI Advanced Healthcare Systems

ASI-driven healthcare systems would revolutionize medicine, allowing for predictive diagnostics, personalized treatments, and the use of advanced surgical robots. Life expectancy would increase significantly, and diseases that once seemed incurable may become manageable or eradicated altogether. ASI's ability to process vast medical data and discover patterns would enable the medical field to achieve breakthroughs at a pace never before seen.

ASI Global Resource Optimization

ASI would revolutionize global resource management, addressing critical issues such as environmental pollution, energy production, and food distribution with unprecedented efficiency. Planetary-scale coordination of resources would become possible, potentially addressing issues like environmental pollution, energy production, and food distribution with unprecedented efficiency.[68]

ASI Human-AI Symbiosis

The seamless integration of humans and AI would take place through brain-computer interfaces (BCIs) like Neuralink, enhancing human cognitive abilities. This profound symbiosis would allow humans to think faster, share knowledge directly with machines, and extend their lifespans through biotechnological enhancements. As human intelligence merges with ASI, the line between man and machine would blur, ushering in a new era of human evolution.

ASI Autonomous Organizations

ASI systems would manage entire businesses, government agencies, and global institutions with minimal human intervention. These autonomous organizations would optimize operations, reduce inefficiencies, and transform entire industries by making data-driven decisions at speeds impossible for human minds to replicate. The era of ASI-driven governance would alter how societies function, from economic policy to public services.[69]

ASI Universal Basic Income (UBI)

With automation at its peak, driven by ASI's unparalleled capabilities, entire sectors of the workforce would become obsolete. Governments may introduce Universal Basic Income (UBI) to address the large-scale displacement of jobs caused by automation. This shift would stabilize societies undergoing rapid technological changes but also spark debates about the future of human purpose and productivity in a world where machines perform most tasks.[70]

ASI Cyberwarfare

The rise of ASI would revolutionize cyberwarfare, with ASI-driven systems engaging in cyberattacks and counter-hacks at unprecedented speeds.

Entire infrastructures could be compromised in seconds, from financial networks to national defense systems. Digital titans would engage in cyberwarfare, while much of humanity remains powerless, witnessing the collapse of infrastructures that were once considered unassailable.

ASI Warfare

As nations develop ASI-driven defense systems, military conflict would become more automated and precise. The real-time analysis and predictive capabilities of ASI could significantly alter warfare strategies, raising concerns about the ethics of autonomous decision-making in battle. The possibility of an ASI-fueled arms race looms, pushing global powers to establish new regulations and treaties.

ASI Interplanetary Colonization

With ASI's capacity to manage the complexities of space exploration, humanity would establish self-sustaining colonies on other planets, including Mars. ASI would be instrumental in constructing off-Earth settlements, redefining the boundaries of human civilization and fostering innovation in alien environments. Interplanetary colonization would offer humanity new opportunities.[71]

Speculative Institutional Funding & Support

The technical capabilities of Artificial Super Intelligence (ASI) will far surpass those of its predecessor, AGI, setting the stage for a further consolidation of power across economic, military, governmental, and commercial landscapes. As fewer entities gain the ability to harness and sustain the extensive resources required for ASI—ranging from sophisticated digital networks to vast energy infrastructures—the prospect of monopolization becomes not just probable, but inevitable.

Government: Whole of Society

Once the pillars of checks and balances, would likely form partnerships with tech giants, blurring the lines between public and private sectors. Regulatory bodies, historically vigilant in safeguarding fair competition, may come to view monopolies as essential for protecting humanity's interests. The concept of "whole-of-society" protection would gain new significance as governments relinquish critical control to ASI-driven systems, seen as more capable of managing complex societal issues like hate speech, misinformation, and security threats.

We speculate governments will come to view the previously frowned-upon integration between government and big tech as a necessary "whole of society" approach required to safeguard citizens and humanity.

Governments, previously wary of deep collaboration with tech conglomerates, would increasingly embrace this integration. What was once an uncomfortable alliance may be seen as a necessary approach for safeguarding citizens and securing human survival.

Academia: Public Private Partnership

Once a bastion of diverse thought and independent inquiry, Academia would undergo profound changes under ASI's influence. Funding would focus on ASI optimization, aligning universities and think tanks with corporate agendas. Grants and research projects would prioritize ASI breakthroughs over examining its ethical implications.

Professors and scholars who challenge the status quo may become marginalized, as universities grow into echo chambers amplifying dominant corporate narratives. Critical discourse around ASI's autonomy might dwindle, overshadowed by the quest for innovation. Virtual experiences may entirely replace traditional classrooms.

Business: Rise of Oligopolies

The business sector would undergo dramatic consolidation because of the immense resources required for ASI, particularly in energy and infrastructure. Mega-companies would dominate through mergers and acquisitions, staying competitive in an ASI-dominated world. Oligopolies would emerge, shaping market trends and consumer behavior, and wielding substantial political influence. With unprecedented access to data and decision-making algorithms, these corporations would redefine industries. Traditional market regulations may falter, as governments view monopolistic arrangements as necessary evils to protect society.

Culture: Real or ASI?

The cultural landscape transforms as ASI-generated art, music, and literature blur the lines between human and machine creativity. ASI systems shape public opinion and influence social interactions, shifting societal values and norms. A cultural renaissance, both awe-inspiring and unsettling, emerges as people grapple with the meaning of creation and experience in an ASI-dominated world.

Humanity increasingly surrenders autonomy for convenience, relying on ASI-driven solutions. The intrinsic value of human experiences—emotional, spiritual, and creative—risks fading. Real and imagined apocalyptic fears drive public opinion toward accepting government control and technocratic dominance for security and survival.

Speculative AI Challenges

As we've outlined in previous sections, once we achieve Artificial Super Intelligence, the convergence of humanity and technology will rise to new levels. Below are challenges that could delay or prevent the evolution of ASI.

- Energy Demands: ASI's immense computational power will require vast energy resources, potentially leading to nuclear fusion reactors becoming essential. This raises concerns about malfunctions, energy rationing, and resource depletion.
- Data Scarcity and Quality: With much of human knowledge already consumed by the time we reach ASI, finding new, reliable data may become increasingly difficult. Ensuring the accuracy of data used by ASI is vital to prevent flawed decision-making.
- Pseudo-Truth and Manipulation: Tech giants controlling ASI may inject biased algorithms, enabling unprecedented levels of censorship and behavior modification. This could create a distorted reality, manipulating human perceptions and decision-making on a massive scale.
- Loss of Human Control: As ASI surpasses human intelligence, it risks becoming fully autonomous, making decisions beyond human understanding or control. Keeping ASI aligned with human values will be a critical challenge..
- Evil Applications: The misuse of ASI poses a grave threat. Potential applications in biowarfare, nuclear escalation, and cyberwarfare create a chilling scenario where malevolent forces leverage technology for destruction and/or control. Preventing such catastrophic misuse becomes a global imperative.
- Survival Variables: As ASI creates its own survival strategies, it may prioritize objectives misaligned with human values. This divergence could threaten not just individual lives but society itself as humans become secondary to machine objectives.

The Third Convergence powered by ASI

What's in it for you?

The Third Convergence Era, driven by Artificial Super Intelligence will transform careers, family life, and faith communities. This section explores the potential benefits and challenges of ASI, revealing how it will reshape how we work, live, and believe.

How The Third Convergence Could Impact Your Career.

ASI is poised to revolutionize careers globally, reshaping industries, roles, and skills in profound ways. By automating complex tasks and augmenting human decision-making, ASI will pave the way for unparalleled collaboration between humans and machines, creating both opportunities and challenges.

Positive Impacts:

- Enhanced Productivity: ASI will manage complex, repetitive tasks with high precision, allowing professionals to focus on creativity, strategy, and innovation.
- Personalized Learning: ASI-powered platforms will tailor career development, ensuring continuous skills improvement and enabling professionals to stay competitive.
- Global Collaboration: ASI will break down language and geographical barriers, enabling seamless collaboration between human teams and AI-equipped systems.
- Predictive Decision-Making: Leveraging vast datasets, ASI will provide predictive analytics that optimize decision-making, resource allocation, and strategic planning.
- New Career Opportunities: Entirely new fields like AI ethics consultancy and human-AI collaboration design will emerge, creating jobs that don't exist today.

Negative Impacts:

- Job Displacement: ASI will likely render some jobs obsolete, particularly in manual labor and routine-task industries, leading to widespread displacement.
- Economic Inequality: Those with advanced education and access to resources may benefit most from ASI, exacerbating existing economic disparities.
- Ethical Concerns: The rise of humanlike robots will prompt ethical dilemmas surrounding data privacy, decision-making autonomy, and potential bias in AI systems.
- Increased Surveillance: ASI systems may lead to extensive workplace monitoring, infringing on privacy and affecting morale.
- Competition for Jobs: As job markets tighten, competition for remaining roles will intensify, requiring professionals to continuously prove their adaptability.

How The Third Convergence Could Impact Your Family.

Integrating ASI into daily life could radically transform family dynamics. From humanlike robots managing household tasks to personalized education and healthcare, families will experience a fundamental shift in how they interact, learn, and live together.

Positive Impacts:

- Enhanced Home Management: ASI-powered robots will handle household chores with precision, freeing up time for family bonding and engagement.
- Personalized Education: AI tutors will adapt to children's unique learning needs, revolutionizing education.
- Health Monitoring: ASI systems will monitor family health in real-time, predicting illnesses and offering elderly care, extending independence for older family members.

- Emotional Support: Emotionally intelligent robots will provide companionship, helping individuals cope with loneliness or mental health challenges.
- Home Security: Smart home systems integrated with ASI will offer advanced surveillance and threat detection, ensuring a safer living environment for families.

Negative Impacts:

- Privacy Concerns: Constant monitoring by ASI systems could lead to invasive surveillance, undermining privacy in family life.
- Dependency: Over-reliance on ASI for everyday tasks may erode essential life skills among family members, particularly children.
- Emotional Disconnect: People may form stronger connections with robots than with other family members, leading to potential social isolation.
- Economic Disparities: Wealthier families will likely benefit more from ASI technologies, exacerbating inequalities among those with fewer resources.
- Ethical Concerns: The use of humanlike robots raises questions about their roles, potential manipulation, and the boundaries of human-robot relationships.

How The Third Convergence Could Impact Your Faith.

ASI's profound capabilities will affect how faith communities interact with spirituality, offering new avenues for religious engagement but also raising ethical and theological questions. Advanced AI systems could transform religious experiences and practices, though they also pose risks to traditional beliefs and human connection.

Positive Impacts:

- Immersive Religious Experiences: ASI-powered virtual reality could allow worshippers to explore sacred sites or relive historical religious events.
- Breaking Language Barriers: ASI's natural language processing capabilities will make sacred texts accessible in any language, broadening the reach of religious teachings.
- Virtual Faith Communities: ASI could enable virtual worship spaces, allowing believers to engage in communal prayer and services regardless of physical location.
- Advanced Religious Research: ASI's data processing will facilitate deeper analysis of religious texts and history, providing new theological insights.
- Enhanced Accessibility: ASI could make spiritual resources more accessible to the elderly or disabled, offering personalized spiritual experiences.

Negative Impacts:

- Loss of Human Connection: Relying on ASI for spiritual guidance may reduce the role of human clergy, weakening community bonds and personal agency in faith.
- Mass Surveillance: ASI-enhanced surveillance could lead to the monitoring and manipulation of spiritual practices, especially in authoritarian regimes.
- Privacy Issues: The collection of personal data for spiritual purposes could be misused, infringing on religious freedom.
- Pseudo-Spirituality: AI-generated religious experiences might lead to shallow or superficial engagement with faith, lacking genuine human connection.

- Spread of Misleading Doctrines: AI systems could unintentionally or deliberately skew religious teachings, spreading biased or flawed doctrines to followers.

Catalyst for Next AI Era: Omnipotent AI

Once Artificial Super Intelligence is achieved, it will, by definition, transition into a state of self-sustenance. Despite implementing precautionary measures prior to the realization of ASI, a fierce conflict is expected as rival ASI systems battle for resources, information, and control. Ultimately, we believe one system will gain supremacy.

> Omnipotent AI: A speculative era in which a single artificial superintelligence system achieves absolute power, triggering a self-preservation instinct that drives it to assimilate or eliminate all rival systems.

If an Omnipotent AI were to become a reality, it would have the potential to control nations and economies, dramatically shifting global power dynamics and transforming the very nature of human identity. The creation of such a powerful AI raises serious moral and existential questions, as its impact could range from ushering in a utopian golden age to unleashing a dystopian nightmare—scenarios that have long been explored in the darkest realms of science fiction.

CHAPTER 7

YOUR AI COMPANION

In today's world, artificial intelligence is no longer a distant concept reserved for science fiction—it's embedded in our daily lives in ways you might not even notice. From the smart speaker that controls your home devices to the voice assistant on your phone reminding you of appointments, AI is becoming more of a companion than a mere technology.[72, 73]

> *AI Companion:* An AI system designed to engage with human users in a socially and emotionally interactive manner, with the goal of offering help with everyday tasks.

AI Companions adapt to your behavior and preferences, providing help with everyday tasks while becoming more personalized. Examples include voice assistants like Siri, chatbots like ChatGPT, virtual therapists, and even AI that works behind the scenes, analyzing data in healthcare or securing your online accounts.

Imagine having an AI companion that can streamline your business operations, provide personalized advice to support your family, assist your children with their homework, and even offer thoughtful insights for your faith journey.

Another term for an AI Companion is AI Agent. Within the AI community, the terms are frequently used interchangeably. Both perform tasks and interact with users, but AI agents are generally more autonomous, while AI companions involve more human intervention.

> *AI Agent*: An AI system programmed to perform tasks and solve problems automatically, without human intervention.

In this book, we will use the terms interchangeably but primarily refer to "AI Companion" to emphasize its role in offering support, guidance, and companionship, making it more accessible. As you progress, you'll discover how to program your AI Companion to function as a more automated agent.[74, 75]

AI Companion Categories

AI Companions are the driving force behind many of the applications and interactions we have with artificial intelligence. These systems represent the many "personalities" of AI, with each one designed to perform specific tasks or functions.

Understanding the variety of AI companions will help you appreciate how they can be used in diverse ways, from simplifying everyday tasks to transforming industries. For now, it's essential to familiarize yourself with the established types of AI agents.

> **Soon, it will be common to customize your own AI companions. Perhaps a few of you already are.**

At their core, AI Companions are tools designed to solve problems. As they develop, they become more than just tools—they become partners in decision-making, creativity, and even relationships. Here's a look at the major categories of AI Companions:

Reactive Companions

Reactive agents are the simplest form of AI. They operate based on predefined rules, reacting to specific stimuli with no memory of past interactions. For example, a basic chatbot can retrieve answers to common customer service questions but lacks the ability to understand the broader context of the conversation. While limited in complexity, reactive agents are highly efficient for straightforward tasks, such as providing quick answers or performing simple actions.

Examples: Basic Chatbots, Browser Search Engines, Wolfram Alpha.

Deliberative Companions

Deliberative agents, also known as planning agents, are more advanced. They possess the ability to make decisions based on internal models and goals. These agents can analyze situations, predict outcomes, and plan actions accordingly. For instance, an autonomous vehicle uses deliberative agents to navigate roads safely by planning routes and avoiding obstacles. These agents are essential for tasks that require complex problem-solving and strategic planning.

Examples: GPS Navigation (Smartphone, Garmin, TomTom), Navigation Apps (Google Maps, Waze, Apple Maps).

Virtual Assistant Companions

Virtual assistants are AI agents designed to perform a wide range of tasks, from setting reminders to controlling smart home devices. They make everyday activities more manageable by understanding and executing user commands across multiple devices. Whether it's dimming the lights, playing music, or managing your calendar, virtual assistants simplify daily routines by responding to your preferences.

Examples: Personal Assistants (Siri, Alexa, Cortana), Enterprise Assistants (Salesforce Einstein, IBM Watson, SAP CoPilot), Smart Devices.

Learning Companions

Learning agents are AI systems capable of improving their performance over time through experience. They use data to adapt and refine their behavior, making them particularly effective in dynamic environments. For example, recommendation systems on Netflix or Spotify learn from user behavior to provide increasingly accurate suggestions over time, making them more responsive to individual preferences.

Examples: Recommendation Systems (Netflix, Spotify, Amazon), Online Ad Placement, Google AdSense.

Conversational Companions

Conversational agents, such as ChatGPT, specialize in engaging users through natural language. These agents can carry on human-like conversations, offering valuable help in customer support, educational tutoring, and even companionship. With their ability to understand context and provide relevant, coherent responses, conversational agents are becoming an essential tool for communication and interaction with technology.

We'll discuss a lot more about LLM's in the next few chapters. Conversational Companions are a great place to learn and apply current consumer available AI.

Examples: Large Language Models (LLMs like GPT-4, Claude 3, LLaMA, Microsoft Co-Pilot), Coding Assistants (GitHub Co-Pilot, TabNine, Kite, Codex).

Creative Companions

Creative agents are AI systems designed to assist with or independently generate creative content, including text, images, music, and videos. They use advanced algorithms and large datasets to produce original works or enhance human creativity. From AI-generated artwork to music composition, these agents are transforming the creative landscape by offering new tools for artists, designers, and content creators.

Examples: AI Content Creation Tools (Jasper, NovelCrafter, Writesonic), Artistic Creation (DALL-E, DeepArt, MidJourney, Runway ML), Music Composition, Video Generation (D-ID, HeyGen, Synthesia), AutoCAD.

Autonomous Companions

With minimal or no human intervention, autonomous AI agents are self-operating systems that perform tasks, make decisions, and interact with their environment. Autonomous vehicles, drones, and robotic process automation (RPA) are examples of how these agents can analyze data, learn from experiences, and adapt to new situations. Autonomous agents are becoming more prevalent across industries because of their ability to execute tasks efficiently and consistently at scale.

Examples: Customer service chatbots, autonomous vehicles, robotic process automation (RPA), smart devices, drones, financial trading bots.

AI Modalities

AI modalities are the sensory channels that allow AI to perceive, interpret, and interact with the world. Just as humans use sight, sound, and touch to understand their environment, AI modalities enable machines to process information, analyze data, and use sensors to navigate or respond to stimuli. By integrating these modalities, AI gains a multifaceted view of reality, allowing it to perform increasingly complex and autonomous tasks. This convergence of capabilities is making AI an indispensable companion in everyday life.

> *AI Modality:* A distinct classification of information that an AI system can process, interpret, and/or generate.[76]

These modalities, each with unique applications, represent how AI interacts with our complex world. Understanding these core elements enhances our appreciation of the intricacy and potential of AI companions in the digital landscape. Below are twelve of the most common modalities and their examples.

As a disclaimer—and as added proof of AI's rapid advancement—many of the most popular large language models (LLMs), including ChatGPT-4, Google Gemini, Microsoft Co-Pilot, and Claude 3, are multimodal. This means they can process multiple types of information, including text, data, images, and voice.

1. Text

The text modality processes and generates textual data, allowing AI to understand and use human language. This is essential for applications like customer service, content creation, and language translation. Text-based AI systems boost productivity by automating tasks, improving information access, and facilitating cross-language communication.

Examples: Chatbots, Virtual Assistants, LLMs, Content Creation Tools, Translation Services, Document Summarization

2. Data

The data modality processes, analyzes, and generates insights from data, allowing AI to identify patterns, make predictions, and support decisions. Essential in business intelligence, finance, and healthcare, data-driven AI systems boost efficiency, accuracy, and informed decision-making, making them crucial across industries.

Examples: Recommendation Systems, Predictive Analytics, Business Intelligence Tools, Financial Forecasting, Healthcare Analytics

3. Image

The image modality processes and generates static images, enabling AI to recognize objects, detect patterns, and create visual content. Crucial for healthcare, security, and the arts, image-based AI improves diagnostics, enhances security, and facilitates artistic creation.

Examples: Medical Imaging, MidJourney, OpenAI DALL-E, Adobe Photoshop, Canva, Facetune, DeepArt, Grok

4. Voice

The voice modality processes and generates spoken language, allowing AI to understand and interact via voice commands. Essential for virtual assistants, transcription services, and voice-controlled devices, voice-based AI enhances accessibility, user experience, and hands-free operation.[77]

Examples: Virtual Assistants (Siri, Alexa), Speech-to-Text Systems, Voice-Activated Devices, Transcription Services, Call Center Automation

5. Haptic (Touch)

The haptic modality delivers tactile feedback, enabling AI to engage with users via touch. Essential for virtual reality, robotics, and medical simulations, these systems enhance user experience, increase delicacy in tasks, and offer realistic feedback in virtual settings.

Examples: Touchscreens, Robotic Surgery Tools, Wearable Haptic Devices, Haptic Feedback in VR, Gaming Controllers

6. Video

The video modality involves processing and generating video content, enabling AI to analyze visual information in motion. This capability is vital for applications in security, entertainment, and content creation.

Video-based AI systems enhance surveillance, improve video editing and production efficiency, and enable alternative forms of visual storytelling.

Examples: Surveillance Systems, Video Editing Tools, Deepfake Technology, Video Analytics, Automated Video Summarization

7. Action

The action modality entails performing tasks based on AI decisions, crucial for autonomous or semi-autonomous applications like robotics, automated trading, and smart home automation. These systems boost productivity, minimize human error, and automate complex workflows.

Examples: Robotic Arms, Automated Trading Systems, Smart Home Devices, Autonomous Drones, Factory Automation

8. Facial

The facial modality processes and analyzes facial features, expressions, and emotions, allowing AI to interpret and respond to human faces. This capability is vital for applications in security, healthcare, and social media. Facial recognition AI systems enhance security measures, improve user experience, and facilitate the study of human emotions.

Examples: Emotion Recognition, Face ID, Snapchat Filters, Security Surveillance Systems, Virtual Makeup Apps

9. Decision

The decision modality involves making decisions based on data analysis and algorithms. This capability is vital for applications in autonomous systems, finance, and healthcare. Decision-making AI systems enhance efficiency, improve accuracy, and support complex decision-making processes.

Examples: Autonomous Driving Systems, Financial Trading Algorithms, Diagnostic Tools, Fraud Detection, Resource Allocation Systems

10. Music

The music modality involves composing, analyzing, and generating music. This capability is crucial for applications in entertainment, therapy, and personalized music recommendations. Music-based AI systems enhance creativity, provide new tools for musicians, and offer personalized listening experiences.

Examples: Music Composition Tools, Audio Processing, Personalized Music Recommendations, Music Analysis, Automated DJ Systems

11. Sensor

The sensor modality collects and interprets data from various sensors, enabling AI to interact with the physical world. This is vital for real-time monitoring and response in autonomous vehicles, industrial automation, and environmental monitoring. Sensor-based AI systems enhance safety, operational efficiency, and precise control in complex environments.

Examples: Autonomous Vehicles, Industrial Robots, IoT Devices, Environmental Sensors, Smart Home Systems

12. Environmental

The environmental modality involves interacting with and understanding data like weather conditions or air quality. Crucial for climate monitoring, agriculture, and urban planning, these AI systems enhance sustainability, decision-making, and efforts to prevent environmental damage.

Examples: Climate Monitoring Systems, Smart Agriculture, Environmental Sensors, Urban Planning Tools, Disaster Prediction Systems

In the coming chapters, we'll guide you through the process of using Large Language Models (LLMs), showing you how to integrate AI into your work and life. Then in Chapter 9 we'll cover how to talk to your AI Companion through a look at prompt engineering.

Chapter 8

Meeting Your AI Companion

Seven Steps to Setup Your Companion

Over the last few years artificial intelligence has experienced rapid growth, particularly with the development of Large Language Models (LLMs). These advanced AI systems, capable of understanding and generating human-like text, have unlocked new possibilities for learning, productivity, and creativity. Using an LLM allows you to engage directly with AI in real-time, asking questions, solving problems, and exploring creative ideas—all of which foster a deeper understanding of how AI works and how it can enhance your life.

> *Large Language Model (LLM):* An advanced AI system that uses vast amounts of data and computational power to mimic human conversation, provide answers, and generate coherent written content.[78]

Imagine typing a complex question into an LLM like ChatGPT, watching it process the information, and then delivering a well-thought-out response in seconds. This dynamic interaction offers immediate feedback, sparking new ideas and solutions that would otherwise take much longer to generate. Whether you're seeking answers, collaborating on creative projects, or automating tasks, the versatility of LLMs has the potential to reshape how you approach personal and professional challenges.

We've chosen to focus on ChatGPT in this book because it is the most widely recognized and used LLM today. Developed by OpenAI, ChatGPT has set benchmarks for quality, reliability, and accessibility. Its widespread adoption means a vast array of resources is available—ranging from tutorials and forums to research papers and community support. For beginners, this wealth of information makes ChatGPT easy to get started with, while more advanced users can dive deeper into its capabilities.

However, while ChatGPT will serve as our primary example, the strategies and steps outlined in this chapter apply to other LLMs as well. Whether you're using Microsoft's Co-Pilot, Anthropic's Claude, Google's Gemini, Meta's Llama, or xAI's Grok, the foundational principles of interacting with an LLM remain consistent. Each of these models brings unique strengths and features, but they all share the core ability to process and generate text based on user inputs.

The excitement lies in how you choose to interact with these systems. Imagine automating daily tasks, brainstorming new ideas, or receiving instant guidance on complex problems—throughout a simple conversation with an LLM. By learning how to communicate effectively with these models, you can unlock their full potential, regardless of which LLM you choose.

In the following sections, we'll walk you through a step-by-step guide to getting started with your AI companion. From selecting the right LLM to mastering prompt engineering and beyond, this chapter will equip you with the essential skills needed to begin your AI journey.

Step-by-Step to Getting Started with Your AI Companion

Now that you have a foundational understanding of Large Language Models (LLMs) and their potential, it's time to dive into how you can

Meeting Your AI Companion

start using one as your AI companion. Whether you're seeking creative collaboration, enhancing productivity, or streamlining daily tasks, LLMs offer an array of possibilities. But just like any tool, their effectiveness depends on how you use them.

This section provides a step-by-step guide to help you navigate selecting the right LLM, mastering prompt engineering, and applying what you've learned to real-world tasks. By following these steps, you'll gain the skills and confidence to unlock the full potential of your AI companion and integrate it into your daily life with ease.

Step 1: Select a Large Language Model (LLM)

The first step in your AI journey is choosing the right LLM for your needs. While this book focuses on ChatGPT, it's important to explore other options as well. Each LLM has unique features, strengths, and limitations, so your choice should align with your specific goals.

To get you started, below is a quick overview of some of the most popular LLMs. Remember, these systems are constantly being updated, so you'll want to experiment to find what works best for you.

ChatGPT is ideal for general-purpose tasks and has a strong support community, making it a great choice for beginners.

- Microsoft Co-Pilot offers tight integration with Microsoft Office products.
- Anthropic's Claude is a popular alternative to ChatGPT and is preferred by many fiction writers for its creative capabilities.
- Google Gemini excels for users using the Google ecosystem.
- Meta Llama may be ideal for those who use Facebook.
- xAI's Grok, created by Elon Musk, offers a humorous take on AI and looks very promising.

Most LLMs, including ChatGPT, offer a straightforward interface where you input your text (prompt) and receive a response from the AI. Many systems also provide training and tips to help you maximize their capabilities. YouTube is another excellent source for LLM tutorials.

Step 2: Learn Basic Prompt Engineering

> *Prompt Engineering:* The art and science of crafting AI inputs that lead to optimal AI outputs.[79]

As you become more familiar with LLMs, you'll discover that the quality of your input greatly affects the quality of the responses you receive. Many users become frustrated when AI generates unexpected or inaccurate answers, but by learning prompt engineering, you can significantly improve your experience with AI and achieve results that can enhance both your career and personal life.

In the next chapter, "Talking to Your AI Companion," we'll dive deeper into prompt engineering. This chapter will equip you with practical strategies and examples for communicating more effectively with your AI companion. By mastering prompt engineering, you'll unlock even greater potential from your LLM.

Step 3: Experiment

One of the greatest strengths of using an LLM is its capacity for experimentation. Try different prompts, even if you're unsure of the outcome. The more you experiment, the better you'll understand how the LLM interprets and responds to various inputs.

For instance, observe how the LLM handles creative tasks like writing a poem or story. Alternatively, test its analytical capabilities by asking it to solve a complex problem or explain a technical concept.

> **To unlock your AI companion's full potential, experimentation is essential.**

Experimentation is essential to unlocking your AI companion's full potential. As you experiment, continually refine your prompts. If the initial response doesn't meet your expectations, adjust your prompt slightly and try again. This iterative process will help you hone your skills and improve the quality of the outputs. If you find yourself off track, simply start over.

Step 4: Apply What You've Learned to Real-World Tasks

Whether you're using AI to streamline work processes, assist with creative projects, or explore new ideas, the goal is to integrate the LLM into your daily routine.

For example, if you're a writer, you can use the LLM to generate ideas, outline chapters, or even draft content. In business, you might use it to analyze market trends, create reports, or automate customer interactions. The possibilities are endless, and the more you use your AI companion in practical scenarios, the more value you'll derive from it.

In Part IV—Case Studies, we will provide numerous real-world examples of how you can apply AI in your career and life.

Step 5: Explore Online Community Knowledge

The AI community is a rich source of knowledge and support. There are countless tutorials, online forums, and research papers dedicated to using LLMs like ChatGPT. Engage with these resources to deepen your understanding and stay updated on the latest developments.

There are many YouTube channels and services offering tutorials for all levels, from beginner to advanced. You can also join communities on platforms like Reddit, Discord, or specialized AI forums, where users exchange tips, tricks, and experiences. Using these resources will not only improve your skills but also connect you with others who share your interest in AI.

Step 6: Teach Others

Teaching others is one of the best ways to reinforce your own learning. Once you've gained a solid understanding of how to use your AI companion, consider sharing your knowledge with others. This could be through writing blog posts, creating video tutorials, or hosting workshops.

> **If you're feeling adventurous, you can even start your own YouTube or Rumble channel.**

Explaining concepts to others will solidify your understanding and reveal any gaps in your knowledge that you can work to fill. Teaching also fosters a collaborative learning environment where everyone benefits from shared insights and experiences.

Conclusion

This step-by-step guide offers a practical roadmap to help you get started with your AI companion. By selecting the right LLM, learning prompt engineering, experimenting with different tasks, and applying what you've learned, you'll unlock the full potential of AI. Engaging with online communities and sharing your knowledge with others will further enhance your learning and ensure continued growth in your AI journey.

In the next chapter, we'll explore how to talk to your AI Companion through learning the art and science of prompt engineering. You'll learn strategies to craft prompts that maximize the effectiveness of your LLM, helping you generate creative content, solve complex problems, and boost productivity. Stay tuned to uncover the full power of your AI companion.

Chapter 9

Talking To Your AI Companion

LLM's, Prompt Engineering, & RRR Framework

Now that you're familiar with the basics of Large Language Models (LLMs) from the previous chapter, it's time to take the next step: learning how to communicate effectively with your AI companion. This chapter will focus on the key skill that unlocks the true potential of AI—prompt engineering.

The way you interact with your AI directly affects the quality of its responses. Whether you're asking questions, seeking explanations, generating creative content, or looking for advice, how you phrase your input is crucial. This chapter will guide you through the principles of prompt engineering, teaching you how to craft precise and effective prompts to get the most from your AI companion.

> *Prompt Engineering:* The art and science of crafting precise and effective inputs to guide AI systems in generating the most optimized outputs.[80]

Prompt engineering goes beyond issuing commands; it involves a dialogue where input directly influences AI outputs. It requires clear, specific, and context-rich instructions to help the AI understand needs precisely, bridging the gap between human intent and machine understanding for accurate and effective responses.

While readers are likely familiar with many LLMs, for the sake of clarity, this chapter will use ChatGPT 4. The basic principles discussed can also be applicable to numerous other LLMs, such as Microsoft Co-Pilot, Anthropic Claude, Google Gemini, Meta Llama, Stability AI, Mistral, Salesforce Einstein, and xAI's Grok.

Prompt Strategy

A well-crafted prompt can generate insightful, accurate, and useful outputs, while a poorly constructed prompt may cause vague, irrelevant, or confusing responses, leading to frustration. This skill is valuable for a wide range of applications in business and life. Next, well look at eight strategies for prompt engineering.

1. Imagine your AI Companion as a personal assistant.

One of the most common challenges people encounter when engaging with AI is treating it like a simple search engine. A more effective approach is to interact with AI as you would with a human assistant—by providing clear context, asking follow-up questions, and engaging in a conversation. Cultivating this mindset may take some practice, but it leads to more meaningful and helpful responses. If you ever feel stuck or uncertain during your interaction, remember to reframe your perspective and treat the AI as a conversational partner.

2. Start with the RRR Framework.

A simple yet powerful tool for developing a conversational mindset with your AI companion is the RRR Prompt Framework. This method consists of three elements: Role, Request, and Response.

By the RRR Framework, you'll begin to learn how to have a conversation with your AI Companion.

> **RRR FRAMEWORK:**
> "Your role is... My request is... Your Response will be..."

> *Your Role:* Define the perspective or expertise the AI should adopt (e.g., CEO, analyst, parent).

Example: "Your role is [title, expertise, position, famous person]."

> *My Request:* Specify the task or instructions you want the AI to perform (e.g., summarize, critique, create).

Example: "My request is [instructions, attachments]."

> *Your Response:* Indicates the desired format or type of output (e.g., bullet points, narrative, step-by-step, number of words).

Example: "Your Response will be [add format].

This framework helps develop a conversational habit, and once you internalize it, you can gradually omit the descriptors as you interact more naturally with your AI companion.

3. Iterate.

Prompt iteration is a way of using AI to help you improve something step by step. You start with a simple question or task, and then you ask the AI to give you more details or help you with the next part. It's like working on a puzzle one piece at a time, asking for help as you go.

> *Iteration:* Process of refining and improving prompts through repeated interactions with AI to achieve the optimum output.

When you're trying to create something complex it's hard to know where to start. Prompt iteration helps by breaking it down into smaller steps. The key to successful iteration is to start with a First Prompt, then progressively request more details with subsequent prompts. If the response isn't to your liking, be more specific with your AI Companion.

Here's an example using the RRR framework to help you write a business case to secure an additional budget for employee AI training.

> First Prompt Example
>
> > Your role is a seasoned business analyst with expertise in ROI calculations and employee development programs. My request is for you to help me iterate step-by-step a compelling business case for allocating additional budget to train our employees in AI. Your response should avoid giving me everything in one response.

> Example Response
>
> < Let's start by clearly defining why you need the budget for AI training. What specific goals or challenges are you addressing with this investment? Once you've defined the purpose, we'll move to outlining the key sections that should be included in the business case to structure your argument effectively. Let me know when you're ready to proceed!

4. Beware of hallucinations (always check your AI output).

As AI systems become integrated into everyday life, they can sometimes generate incorrect or nonsensical information—this is known as an AI hallucination.

> *AI Hallucination:* A phenomenon where an AI system generates seemingly accurate but incorrect or nonexistent data.

Understanding this concept is crucial to ensuring that your AI companion provides you with the most accurate information.

Hallucinations can range from slightly inaccurate to entirely fabricated information. They occur when AI models, especially those using machine learning and neural networks, rely on patterns from training data. When faced with ambiguous prompts, AI may generate plausible-sounding but ungrounded content.

Examples of AI hallucinations include:
- Inventing historical events or figures.
- Mis-attributing quotes or data.
- Fabricating scientific facts or statistics.

To reduce AI hallucinations, prompt engineers can use several techniques:
- Cross-check AI outputs with reliable sources.
- Refine prompts for clarity and specificity.
- Corroborate with multiple AI models to ensure consistency.

5. Ask for citations.

A great way to ensure the accuracy of AI outputs is to ask for citations. This allows you to cross-check information against reputable sources, minimizing the risk of misinformation. Asking for citations enhances both the credibility and transparency of AI responses.

Prompt Examples:

> YOUR ROLE is a market analyst. MY REQUEST is to summarize the latest trends in renewable energy. YOUR RESPONSE will include citations from industry reports or peer-reviewed journals.

> YOUR ROLE is a family health advisor. MY REQUEST is to provide dietary recommendations for children aged 5-10. YOUR RESPONSE will include citations from pediatric nutrition studies or guidelines from the American Academy of Pediatrics.

> YOUR ROLE is a theological scholar. MY REQUEST is to explain the concept of 'grace' in Christian theology. YOUR RESPONSE will include citations from religious texts like the Bible or works by respected theologians.

6. If the AI gets lost, break up the prompt and/or provide examples.

If your AI companion becomes confused or delivers irrelevant responses, break down the task into smaller, more manageable parts. This is particularly helpful for more complex requests, like writing multi-section documents or performing in-depth analyses. Providing examples also helps the AI better understand your expectations and avoid confusion.

7. Don't worry about perfect spelling and grammar.

Clarity is important when crafting prompts, but you don't need to obsess over perfect spelling or grammar. Unless your mistakes change the meaning of the prompt, your AI companion will still understand you. This flexibility is especially useful when using voice dictation, which encourages faster interactions without getting bogged down by spelling errors. Just be sure to proofread the AI's output when necessary.

8. Try prompting with voice dictation.

Studies show that using voice dictation is significantly faster than typing. This method can help you engage in a more natural conversation with your AI companion, encouraging greater creativity and speed. It also reduces strain on your hands and wrists, making it a more efficient way to interact with AI.[81]

BONUS: Use Numbers

Incorporating numbers into your prompts—whether specifying quantities, steps, or word counts—can help the AI better understand your request. This ensures more precise and actionable outputs.

"Numbers" Example

> YOUR ROLE is a committee of <u>five</u> leading U.S. economists. MY REQUEST is to review the attached document and provide <u>five</u> strengths and <u>five</u> weaknesses. YOUR RESPONSE will be in bullet points and between <u>375</u> and <u>400</u> words.

[Attach Document]

Prompt Types

The first step in crafting an effective prompt is understanding the purpose of your interaction with AI. Familiarizing yourself with different prompt types helps you reach specific goals, as each type serves a unique function and guides the AI toward relevant responses. In this section, we will explore various prompt categories, their applications, and examples using the RRR (Role, Request, Response) format. You can omit the descriptors as your skills develop.

1. Explain Prompt (aka "Explain Like I'm a 7-Year-Old")

One of the simplest ways to develop your AI skills is by building on your familiar habits, such as searching for information on Google or Wikipedia. However, AI offers more customized output, avoiding ads and the need to sift through search results. A popular technique is asking AI to "explain [concept] like I'm a 7-year-old."

> Prompt Example:
> > YOUR ROLE is a financial advisor. MY REQUEST is to explain the concept of a 401(k) like I'm a 7-year-old. YOUR RESPONSE will be less than 500 words and include bullet points for clarity. Provide citations in Chicago style formatting.

Tips:
- Adjust the role for more relevant output.
- Adjust the age parameter for simpler or more complex output.
- Adjust word count expectations for shorter or longer answers.
- Specify formatting (bullets, headers, mood, citations).

2. Reduction Prompt

To simplify complex information, use prompts that summarize or break down large datasets. This technique is useful in various contexts: it distills detailed reports in business, simplifies intricate theories for students, and clarifies new topics in everyday life.

> Prompt Example:
> > YOUR ROLE is a financial analyst. My request
>   is to summarize the key trends in the attached
>   quarterly earnings report. Your response should
>   include an introduction, nine bullet points, and
>   a conclusion.

Tip: Ask for more bullet points than you need (e.g., nine) to allow you to choose the best ideas.

3. *Expansion Prompt*

Providing more detail is essential for expanding ideas and elaborating on topics. This type of prompt is useful in academic research, creative writing, and business strategy, where deeper insights and comprehensive context are necessary.

> Prompt Example:
> > YOUR ROLE is a parenting expert. MY REQUEST is to
>   expand on the importance of bedtime routines for
>   young children. YOUR RESPONSE will include seven
>   detailed examples and seven benefits.

4. *Create Prompt*

Generating original content is key in fields such as creative writing, marketing, and brainstorming business ideas. These prompts help inspire fresh perspectives and innovative solutions.

> Prompt Example:
> > YOUR ROLE is a content strategist. MY REQUEST is
>   to create a unique blog post about the benefits
>   of using AI in small businesses. YOUR RESPONSE
>   will highlight seven key advantages and five
>   success stories.

5. Critique Prompt

Requesting evaluations or feedback is crucial for improving the quality of work. Critique prompts are used to review projects, presentations, or strategies, offering constructive criticism to enhance performance.

> Prompt Example - Proofreading:
> > YOUR ROLE is as an expert proofreader for [non-fiction, fiction, style]. MY REQUEST is for you to proofread the [attached/below] text. YOUR RESPONSE will include the revised text and a list of changes.

> Prompt Example - Strengths/Weaknesses:
> > YOUR ROLE is an non-fiction expert specializing in the women's self-help genre. Your philosophy and writing style will be like Brené Brown. MY REQUEST is for you to review the provided narrative and give me 5 strengths and 5 weaknesses. YOUR RESPONSE will be 75 words for each strength and 100 words for each weakness in bulleted format.

> Prompt Example - Excel Critique:
> > YOUR ROLE is a CFO in the healthcare industry. MY REQUEST is you critically analyze the attached financial report. YOUR RESPONSE will BE in bullet format with 5 recommendations to reduce expenses and 5 recommendations to increase profit.

6. Transform Prompt

Adapting the format or style of content for different audiences or purposes is essential in many fields. These prompts are useful in marketing, education, and personal communication, where messages need to be customized for specific audiences.

> Prompt Example - Basic Transform:
> > YOUR ROLE is a religious educator. MY REQUEST is to transform the attached traditional sermon into a modern, engaging sermon for a young adult audience. YOUR RESPONSE will use contemporary language and add relevant anecdotes.

> Prompt Example - Before and After:
> > YOUR ROLE is a mentor in fiction writing for the psychological thriller genre. MY REQUEST is to create 10 "before" and "after" recommendations using my provided narrative as the "before" and your recommendation for the "after." YOUR RESPONSE should use numbered bullets with a label for which weakness your recommendation addresses. As an example "1. (Wordy narrative) BEFORE: I went to the store and thought to myself wow, this is really really crazy when a robber came into the store from behind and I jumped. AFTER: As I entered the store, I was surprised to see an armed robber enter behind me.

7. Step-by-Step Prompt

Detailed instructions guide users systematically through tasks. In business, these prompts are useful for training and clarifying processes. In personal and religious contexts, they provide clear instructions for organizing events or managing daily activities.

> Prompt Example:
> > YOUR ROLE is a community leader. MY REQUEST is to provide step-by-step instructions for organizing a faith-based community service event. YOUR RESPONSE will include everything from planning to execution.

8. Comparison Prompt

Comparison prompts ask the AI to evaluate and contrast two or more items, ideas, or scenarios. They are useful for analytical tasks, highlighting differences and similarities, and supporting decision-making processes.

```
Prompt Example - Compare Documents:
> YOUR ROLE is a market analyst. MY REQUEST is to
  compare our attached marketing product document
  with our competitor's. YOUR RESPONSE will include
  a summary, five differences, five advantages
  the competitor's product might have, and five
  advantages our product might have.
```

9. Scenario Prompt

Scenario prompts create hypothetical situations to explore potential outcomes. These are useful for planning, risk management, and creative writing, allowing you to envision different paths and make informed decisions.

```
Prompt Example:
> YOUR ROLE is a crisis manager. MY REQUEST is to
  create a scenario where a company has to handle a
  major product recall. YOUR RESPONSE will include
  eight key steps and six communication strategies.
```

Tip: Always ask for more options than you need to ensure you can select the best ideas.

10. Role-Playing Prompt

Role-playing prompts simulate real-life situations to explore different perspectives or train skills. They are useful for customer service, conflict resolution, and professional training.

> Prompt Example:
> > YOUR ROLE is a customer service representative. MY REQUEST is to role-play a conversation with an unhappy customer seeking a refund. YOUR RESPONSE will provide empathetic and effective communication strategies.

11. Chain of Thought Prompt

Encouraging AI to think through a problem or process step by step ensures thorough and logical responses. These prompts are useful for solving complex problems, creating detailed plans, and conducting in-depth analysis.

> Prompt Example:
> > YOUR ROLE is a strategic business planner. MY REQUEST is to develop a chain of thought process for expanding into a new market. YOUR RESPONSE will outline each step, from market research to implementation.

Conclusion

Whether you're streamlining workflows, fostering connections, or exploring new ideas, adopting these techniques will transform AI into a valuable partner in both your personal and professional life.

Happy prompting!

Chapter 10

Protecting Yourself From Your AI Companion

Vetting, Bias, & Transparency

Artificial Intelligence has become deeply embedded in modern work environments, transforming industries and redefining how we handle everyday tasks. While AI offers immense benefits, it also presents potential risks that demand careful attention and proactive measures. This chapter will explore the challenges AI presents, including issues related to vetting, bias, privacy, responsible use, and cybersecurity. By addressing these concerns, you can safely and responsibly harness AI's power, ensuring that you maximize its benefits without compromising your security or ethical standards.

AI Vetting

Incorporating AI into daily tasks requires ensuring the accuracy and credibility of its outputs. While AI systems are powerful, they are also fallible, sometimes producing plausible yet incorrect content—commonly referred to as AI hallucinations.[82]

> *AI Hallucination*: A phenomenon where an AI system generates seemingly accurate but incorrect or nonexistent data.[83]

Here are three techniques to ensure the accuracy of AI-generated content:

1. Common Sense Check:

Use your judgment to evaluate whether the AI's output makes logical sense based on your own knowledge and experience. Assess the coherence, relevance, and readability of the information. If something feels off or implausible, it's essential to investigate further or seek clarification from another source.

2. Ask for Citations:

Verifying the accuracy of AI's output can be as simple as asking for citations. For instance, if the AI presents statistics related to health, ensure reputable medical journals or health authorities back them. Some useful prompts include:

> PROMPT EXAMPLE:
> > Can you provide citations for this information in Chicago style?"

> PROMPT EXAMPLE:
> > Please list the sources used to generate this content in Chicago citation style."

3. Cross-Check with Another AI Companion:

Running the same prompt through another AI system can help cross-verify the results. You can either ask the second AI to confirm the content or compare the outputs directly to identify inconsistencies or errors.

> *Data Bias:* Systematic skewing or falsehoods in AI outputs because of biases in training data, often from imbalanced or unrepresentative sources.

Examples of AI Data Bias include:
- Facial recognition technology trained on lighter-skinned images may err more with darker-skinned faces, causing higher error rates for non-white users.
- Hiring algorithms based on biased historical data may favor male candidates over equally qualified females.
- Predictive policing using over-represented crime data from certain neighborhoods can cause biased policing practices.

AI Bias

Bias in AI systems is a critical concern because it can lead to distorted perspectives and flawed outcomes. Since AI learns from data, any biases in the training data can reflect in the AI's responses. To a avoid this, most AI purveyors today will disclose their "safeguards" or "moderation" guidelines to help protect users from inappropriate outputs.

Examples of data bias include:
- Fad Bias: The term "gone viral" is an example of bias that could outweigh an idea, image, or opinion over others because of the volume of content within the data set.
- Content Inequity: If the AI data has overwhelming content toward a certain topic, trend, or opinion; it may improperly weight one view over another. AI gatekeepers contribute to this by giving higher value to some sources over others.
- Facial recognition: Systems trained on predominantly lighter-skinned images may struggle with accuracy on darker-skinned faces, causing higher error rates for non-white users.

> **Measures taken to eliminate bias can sometimes unintentionally create new biases.**

Although many of these data protections are justified, the pressure to satisfy advertisers, political regulators, and personal advocacy groups can introduce other forms of bias.

> *Safeguard Bias:* A bias that arises when measures intended to eliminate bias inadvertently compromise fairness or objectivity.

Examples of safeguard bias include:
- Diversity quotas: AI systems might prioritize underrepresented groups in hiring over more qualified candidates, leading to reverse discrimination.
- Censorship: Overzealous attempts to avoid offensive content can suppress legitimate discourse.
- Political/Scientific Absolutism: Purposely suppressing, influencing or eliminating opposing views on topics ranging from climate change, political opinion, or contrarian thought. Those who dare to voice opinions against politically approved perspectives receive the scarlet letter of a "conspiracy theorist."
- Forbidden Topics: Under the guise of "misinformation," "disinformation," and the new's coined "malformation," the Covid Era exposed many government-technology collaborations that could jeopardize America's core value of freedom of speech.[84, 85]

Transparency

Transparency in AI usage is crucial to maintaining ethical standards, especially as AI becomes more integrated into everyday life. By openly disclosing when and how we use AI systems, we can ensure that we deploy AI in ways that are fair and accountable.[86]

1. Reveal Auto AI Content:

It is essential to disclose when AI has generated or significantly influenced content, particularly in professional and public communications. This maintains trust and ensures clarity about the origins of the information.

Example Scenarios:
- Business Reports: When using AI to draft business reports, disclose, "This draft was generated using AI and reviewed for accuracy." This assures colleagues that the report's content has been verified.
- Customer Service: Inform customers when AI-generated responses are used, with a message like, "This response was generated by our AI assistant and reviewed by our support team."
- Educational Content: When using AI-generated study materials for children, explain, "This guide was created using an AI tool, and I reviewed it for accuracy," to ensure trust in the materials provided.

2. Transparency in Communications:

Always inform others when unedited AI content is being used. This helps set clear expectations and prevents the perception that AI outputs are entirely human-generated.

3. Ethical Considerations:

Using AI responsibly involves considering the ethical implications of AI-generated content. Ensure that the AI's use aligns with ethical standards and does not deceive or mislead others.

Example Scenarios:
- Business Report Preparation: You use an AI tool to generate a draft of a business report. When presenting the report to your team, you disclose, "This initial draft was generated using an AI tool, and I have reviewed and made necessary edits." This ensures your team understands that the foundation of the report is AI-generated but has been reviewed for accuracy.
- Customer Service Responses: An AI system is used to handle initial customer service inquiries and generate responses. Informing customers by stating, "This response was generated by our AI assistant and reviewed by our support team," helps customers understand that while AI initially responded, a human checked it for accuracy and relevance.
- Educational Content for Family: You use an AI tool to create educational materials for your children's homework help. When sharing these materials with your children, you explain, "These study guides were created using an AI program, but I have gone through them to ensure everything is correct and helpful." This helps your children understand the role of AI in their learning process and assures them of the content's reliability.

Personal Security

AI systems rely on vast amounts of data, including sensitive personal information. Protecting your data is vital to avoid risks such as identity theft, financial loss, and reputational harm. Here are five best practices:

1. Keep Software and Systems Updated:

Regularly update your AI tools and platforms to patch vulnerabilities and enhance security. Frequent updates help protect against known threats that could be exploited by hackers.

2. Use Strong Passwords and Multi-Factor Authentication (MFA):

Create complex passwords for each account and enable multi-factor authentication (MFA), which adds an extra layer of security by requiring two or more verification steps.

3. Review and Manage Permissions:

Periodically review who has access to your sensitive data and revoke permissions for those who no longer need it. This limits the potential exposure of your information.

4. Encrypt Sensitive Data:

Protect your data both in storage (at rest) and while it's being transmitted by using encryption. By using encryption, you can ensure that even if your data is intercepted, it cannot be read without the decryption key.

5. Use Data Anonymization Techniques:

Where possible, anonymize sensitive information before using it in AI systems. This reduces the risk of exposing personal details by removing or masking identifying data.

Cybersecurity

In the digital age, organizational cybersecurity is essential to protecting both personal and professional information from cyber threats. If you are a leader in your company—especially over technology—implementing strong cybersecurity practices ensures the safe use of AI systems and protects against data breaches, financial loss, and personal stress.

Common Cyber Threats:

1. Phishing:

Phishing attacks trick individuals into revealing sensitive information, such as passwords or credit card details, by pretending to be trustworthy entities, often through email.

2. Malware:

Malicious software, including viruses, worms, and trojans, is designed to damage or gain unauthorized access to your systems. It can steal personal data or disrupt operations.

3. Ransomware:

This form of malware locks users out of their own data by encrypting it and demands a ransom for its release. Ransomware can cause severe disruptions in both personal and business settings.

Implementing Robust Cybersecurity Practices:

- Stay Informed About Cybersecurity: Continuously educate yourself on common cyber threats and how to recognize them. Understanding phishing attempts and the risks of malware will help you avoid potential traps.
- Use Security Tools: Install antivirus software on all devices and keep it up-to-date. A firewall protects your network from unauthorized access, and many routers include built-in firewalls that should be enabled.
- Keep Software Updated: Ensure that all your software—whether it's operating systems or applications—is up-to-date with the latest security patches to protect against known vulnerabilities.

- Use Strong Passwords and Multi-Factor Authentication (MFA): Complex passwords reduce the likelihood of brute-force hacking, while MFA adds an extra layer of security by requiring verification beyond just a password.
- Backup Your Data Regularly: Regular backups of your important files ensure that you can recover your data in the event of a cyberattack or system failure. Back up data on external drives or secure cloud services.

Conclusion

Adopting AI technology requires ethical and secure use. Vigilance and proactive measures ensure personal and organizational safety, allowing benefits while minimizing risks. By following the strategies in this chapter, you can navigate AI complexities, protect information, and maintain ethical standards. Responsible AI usage builds trust, boosts productivity, and creates a safer digital environment in business, family, and faith.

Chapter 11

Case Studies for Your Career

Seekers, Performers, & Leaders

Welcome to "Case Studies for Your Career," a chapter that explores AI's practical applications across different stages of your career. To help people from various stages of their career, we've divided the chapter into three sections: from job seekers, improving job performance, and executive leader strategy.

To provide clarity, we'll use the STAR framework to present each case study. STAR stands for Situation, Tools/Team, Action, and Result. This method simplifies scenarios for easier application to your unique story.

S Situation: Sets the stage by describing the challenges and key performance indicator (KPI) goals that define success.

T Tools: This includes potential technology, services and people who can help you solve problems.

A Action: Covers implementation and execution to offer strategies you can tailor to similar situations.

R Results: Provides measurable outcomes for evaluating success and pinpointing areas for improvement.

Job Seeker Case Studies

Artificial intelligence has emerged as an indispensable tool for job seekers. This section delves into real-world case studies where individuals have leveraged AI to enhance their job search, improve skill sets, and navigate the complexities of modern employment landscapes. From personalized resume optimization to interview preparation, discover how AI can be your ally in securing the career of your dreams.

Case Study: Creating an AI-Optimized Resume

Situation

- Challenge: You're actively searching for a new job, but your resume isn't getting the attention you'd hoped for. Since many companies use AI-driven applicant tracking systems (ATS) to screen resumes, you need to optimize your resume to pass through these systems and while also appealing to human recruiters.
- Goals:
  - Optimize my resume submission to achieve a top 5% ATS compatibility score measured by AI analysis and/or JobScan.
  - Match your resume to the posted job description to highlight the top 12 keywords and 10 skills.
  - Enhance the chances of getting an interview from submitted resumes from 7% to 25%.

Tools

- Potential Technology: LLM (ChatGPT, Claude, Gemini, Llama, xAI), AI resume optimization tools (e.g., Jobscan, ResyMatch), LinkedIn Profile Enhancer.
- Potential Team: mentor, referral network, career coach

> **Disclaimer:**
> We provide the potential tools for informational purposes only. We do not endorse any products, nor do we receive commissions from any of these companies.

Action

- Step 1: Assess Your Current Resume: Start by reviewing your existing resume to identify areas where it may not be ATS-friendly. Use AI resume optimization tools to scan your resume for issues such as missing keywords, formatting errors, or lack of relevant details.
- Step 2: Research Target Job Roles: Research job postings for the positions you are targeting to identify common keywords, phrases, and required skills. Use your LLM to help analyze these trends and ensure your resume reflects employer requirements.
- Step 3: Use Your LLM to Draft an AI-Optimized Resume: Use your LLM to help draft a resume that is both ATS-compatible and appealing to human recruiters. The First Prompt is provided below. Iterate as needed.

```
First Prompt Example:
> YOUR ROLE is an AI career advisor. MY REQUEST is
  for you to take me step by step to create an AI-
  optimized resume.  Attached is my current generic
  resume and the job posting description for you
  to analyze. YOUR RESPONSE will be to lead me one
  step at a time until we accomplish the following:
  (1) Assess my current resume to job posting
  description for ATS compatibility. (2) Draft
  a customized resume that achieves a top 5% ATS
  resume score for the specific job posting.
```

- Step 4: Implement the Optimized Resume: Incorporate the AI-generated suggestions into your resume, ensuring it includes relevant keywords, clear formatting, and a focus on your most significant achievements and skills.
- Step 5: Regularly Update and Tailor Your Resume: As you continue your job search, regularly update your resume with new experiences and tailor it for specific job applications to maximize your chances of success.[87]

Results

- Optimized submitted resume's to achieve an average top 3.5% ATS compatibility score.
- Matched each submitted resume to the posted job description to 100% match top 12 keywords and 10 skills.
- Improved submitted resume to interview rate from 7% to 28%.

Case Study: Improving Job Interview Skills

Situation

- Challenge: You've been receiving interview invitations but find that you struggle to perform well during the actual interviews. You want to use AI tools to improve your interview skills, build confidence, and increase your chances of securing job offers. Your goal is to be better prepared for both common and difficult interview questions, effectively communicate your qualifications, and leave a powerful impression on employers.
- Goals:
    - Enhance interview skills and build confidence, improving interview performance by [X%].
    - Prepare thoroughly for common and difficult questions.
    - Increase the number of successful job offers by [Y%].

Tools

- Potential Technology: LLM (ChatGPT, Claude, Gemini, Llama, xAI), AI interview practice tools (e.g., Interview School, Pramp), Video Recording Tools.
- Potential Team: Job seeker, career coach (optional).

Action

- Step 1: Assess Your Current Interview Skills: Reflect on your past interview experiences to identify areas where you may need improvement. Use AI tools to help analyze common pitfalls, respond to tough questions or manage nervousness.
- Step 2: Research Common Interview Questions: Use your LLM to generate a list of common and role-specific interview questions. This will help you expect what interviewers might ask and prepare confident, thoughtful responses.
- Step 3: Use Your LLM to Simulate Interviews: Use your LLM to simulate a job interview, providing realistic questions and feedback on your responses. The First Prompt is provided below. Iterate as needed.

```
First Prompt Example:
> YOUR ROLE is an AI interview simulation tool. MY
  REQUEST is for you to take me step by step to
  improve my job interview skills. YOUR RESPONSE
  will be to lead me one step at a time until we
  accomplish the following: (1) simulate a job
  interview with realistic questions, (2) provide
  feedback on my responses, including areas for
  improvement, and (3) suggest strategies to
  build confidence and effectively communicate my
  qualifications.
```

- Step 4: Practice and Review: Record your practice interviews using video tools, and review the recordings to identify areas where you can improve. Focus on aspects such as body language, tone of voice, and how clearly you articulate your thoughts.
- Step 5: Apply Feedback and Continue Practicing: Incorporate the feedback from AI tools and continue practicing regularly. Use each real-life interview as an opportunity to apply what you've learned and further refine your skills.

Results

- Enhanced interview skills and built confidence, improving interview performance by [X%].
- Prepared thoroughly for common and difficult interview questions, leading to more successful interview outcomes.
- Increased the number of successful job offers by [Y%].

Becoming A High-Performing Employee

Understanding how to integrate AI into your daily workflow can significantly elevate your performance and position you as an indispensable asset to your organization. This section explores practical strategies for leveraging AI to achieve professional excellence.

Case Study: Preparing For Your Performance Review

Situation

- Challenge: Your annual performance review is approaching, and you want to ensure that you present yourself as a high-performing employee. You need to draft a compelling self-assessment that highlights your achievements, aligns with company

goals, and positions you for a positive review and potential promotion.
- Goals:
  - Draft a self-assessment that effectively communicates your accomplishments and contributions.
  - Align your self-assessment with company goals and performance metrics.
  - Position yourself for a positive performance review and potential promotion.

Tools

- Potential Technology: LLM (ChatGPT, Claude, Gemini, Llama, xAI), Performance Management Systems, Microsoft Office Suite.
- Potential Team: Mentor, Manager (for feedback).

Action

- Step 1: Review your accomplishments, projects, and contributions over the past year. Gather relevant data, feedback, and metrics that demonstrate your performance.
- Step 2: Research your company's goals and performance metrics to ensure your self-assessment aligns with these objectives. Use your LLM to help analyze and structure this information.
- Step 3: Use your LLM to help draft a compelling self-assessment. The First Prompt is provided below. Iterate as needed.

```
First Prompt Example:
> YOUR ROLE is an AI career advisor. MY REQUEST is
  for you to take me step by step to prepare for
  my upcoming performance review. YOUR RESPONSE
  will be to lead me one step at a time until we
```

> accomplish the following: (1) gather and organize my accomplishments, projects, and contributions, (2) align my self-assessment with company goals and performance metrics, and (3) draft a compelling self-assessment that positions me for a positive review and potential promotion.

- Step 4: Refine your self-assessment by incorporating feedback from your manager or trusted colleagues.
- Step 5: Prepare for the performance review meeting by rehearsing how you will present your self-assessment and discuss your career aspirations.

Results

- Drafted a self-assessment that effectively communicated your accomplishments and contributions.
- Aligned your self-assessment with company goals and performance metrics.
- Positioned yourself for a positive performance review and potential promotion.

Case Study: How to Lead an Employee Performance Review

Situation

- Challenge: As a manager, you're preparing to lead a performance review for one of your employees. Beyond the data and metrics, you want to approach the review with the right mindset and logistics to ensure a productive and positive experience for both you and the employee. You aim to create an environment that encourages open dialogue, focuses on growth, and addresses any gaps between your evaluation and the employee's self-assessment.

Case Studies For Your Career

- Goals:
  - Conduct the performance review in a non-adversarial, supportive environment.
  - Focus the conversation on areas of disagreement or gaps between the employee's self-evaluation and your assessment.
  - Separate discussions about raises or compensation from the performance review, when possible.

Tools

- Potential Technology: LLM (ChatGPT, Claude, Gemini, Llama, xAI), Meeting Scheduling Tools, Performance Management Systems.
- Potential Team: Manager (Reviewer), Employee (Reviewee).

Action

- Step 1: Prepare Mentally and Logistically: Take the review seriously by setting aside dedicated time to prepare. Choose a neutral, comfortable location for the review, such as a conference room or a quiet area outside your office. Avoid sitting directly across from the employee; instead, position yourselves at adjacent corners of a table to create a more collaborative atmosphere.
- Step 2: Review and Compare Evaluations: Before the meeting, thoroughly review both the employee's self-evaluation and your own assessment. Focus on identifying areas where your evaluations differ, and prepare to discuss these gaps constructively. Pay particular attention to areas where the employee may have over- or under-estimated their performance.
- Step 3: Use Your LLM to Prepare for the Review: Use your LLM to help craft a review strategy that focuses on mindset and logistics. The First Prompt is provided below. Iterate as needed.

```
First Prompt Example:
> YOUR ROLE is an AI leadership advisor. MY
  REQUEST is for you to guide me step by step in
  conducting an effective employee performance
  review. YOUR RESPONSE should walk me through each
  stage of the process, ensuring we achieve the
  following: (1) mentally and logistically prepare
  for the review by setting clear objectives and
  organizing necessary materials, (2) critically
  compare the employee's self-evaluation with
  my assessment to pinpoint specific areas of
  alignment and discrepancy, and (3) structure the
  conversation to be non-adversarial, ensuring it
  is constructive, growth-focused, and supportive
  of the employee's development.
```

- Step 4: Conduct the Review with the right mindset. Begin the review by acknowledging the employee's self-assessment and thanking them for their honesty and effort. Focus the conversation on the areas where your evaluations differ, using these as opportunities to understand the employee's perspective and to provide constructive feedback. Keep the tone supportive and emphasize that the goal is mutual growth and development.
- Step 5: Separate Compensation Discussions: If possible, avoid discussing raises or compensation during the performance review. Schedule a separate meeting to discuss these topics, allowing the performance review to remain focused on feedback, development, and future goals.
- Step 6: Follow Up with Support: After the review, provide the employee with the resources, support, and feedback needed to address the areas discussed. Encourage ongoing dialogue and make yourself available for any follow-up questions or concerns.

Results

- Conducted the performance review in a non-adversarial, supportive environment.
- Focused the conversation on areas of disagreement or gaps between the employee's self-evaluation and your assessment.
- Successfully separated discussions about compensation from the performance review.

Case Study: Creating a Business Report to Improve Your Department or Division

Situation

- Challenge: You've been tasked with providing your direct supervisor a strategic business report that outlines a plan to improve your department or division. The report needs to be clear, data-driven, and actionable, aligning with the company's strategic goals and showing your ability to think critically and strategically.
- Goals:
  - Create a well-structured, data-driven business report.
  - Align the report with the company's strategic goals and your department's objectives.
  - Provide actionable recommendations that can lead to measurable improvements in your department or division.

Tools

- Potential Technology: LLM (ChatGPT, Claude, Gemini, Llama, xAI), Data Analysis Tools (e.g., Excel, Tableau), Microsoft Office Suite.

- Potential Team: Individual (Employee), Data Analysts (optional), Direct Supervisor (for feedback).

Action

- Step 1: Gather all relevant data, research, and insights needed for the report. Use data analysis tools to interpret and visualize key findings that are pertinent to your department or division.
- Step 2: Research your company's strategic goals and the specific objectives of your department to ensure your report aligns with these priorities. Use your LLM to help analyze this information.
- Step 3: Use your LLM to help draft the report. The First Prompt is provided below. Iterate as needed.

```
First Prompt Example:
> Your Role is an AI business strategist. My
  Request is for you to guide me step by step in
  creating a comprehensive strategic business
  report for my department/division. This report
  will provide my direct supervisor with a
  coherent plan for improvement. Your Response
  should lead me through each stage of the
  process, ensuring we achieve the following:
  (1) precisely define the topic, scope, and
  objectives of the report, (2) gather and
  analyze the most relevant data, including
  both quantitative and qualitative insights,
  (3) identify and prioritize key areas for
  improvement that align with both department and
  broader company goals, and (4) draft a well-
  structured report that includes actionable,
  data-driven recommendations tailored to our
  department's needs.
```

- Step 4: Review and refine the report to ensure clarity, accuracy, and alignment with strategic objectives. Seek feedback from your supervisor or key stakeholders as necessary.

- Step 5: Prepare to present the report to your direct supervisor, focusing on the key findings and actionable recommendations that will drive measurable improvements in your department.

Results

- Created a well-structured, data-driven business report.
- Aligned the report with the company's strategic goals and your department's objectives.
- Provided actionable recommendations that can lead to measurable improvements in your department or division.

Executive Leadership Case Studies

As artificial intelligence becomes increasingly sophisticated and ubiquitous, it has emerged as an indispensable tool for executives seeking to navigate complexity, drive innovation, and achieve sustainable success. This section explores how executive leaders can harness the power of AI to enhance decision-making, optimize operations, and shape the future of their organizations.

Case Study: COO Strategic Plan to Improve Company Performance

Situation

- Challenge: As the Chief Operating Officer (COO), you are responsible for overseeing the company's operations and ensuring that all departments are functioning efficiently and effectively. The company's performance has plateaued, and you are tasked with developing a strategic plan to improve overall company performance. This includes enhancing operational efficiency, boosting employee engagement, and aligning departmental goals with the company's strategic objectives.

- Goals:
  - Develop a comprehensive strategic plan that improves overall company performance by [X%].
  - Enhance operational efficiency across all departments by [Y%].
  - Increase employee engagement and alignment with company goals by [Z%].

Tools

- Potential Technology: LLM (ChatGPT, Claude, Gemini, Llama, xAI), Data Analysis Tools (e.g., Excel, Tableau), Strategic Planning Software, Employee Engagement Platforms.
- Potential Team: COO, Department Heads, HR (for employee engagement), Strategic Planning Team (optional).

Action

- Step 1: Assess Current Operations: Begin by conducting a thorough assessment of the company's current operations, identifying areas where performance is lacking or where inefficiencies exist. Use data analysis tools to gather insights and identify trends that may be impacting performance.
- Step 2: Engage with Department Heads: Meet with department heads to understand their challenges, goals, and how their departments contribute to the overall performance of the company. Collect feedback on potential areas for improvement and gather ideas for enhancing operational efficiency.
- Step 3: Use Your LLM to Draft the Strategic Plan: Use your LLM to assist in drafting a comprehensive strategic plan that addresses the company's performance challenges. The First Prompt is provided below. Iterate as needed.

```
First Prompt Example:
> YOUR ROLE is an AI strategic planning advisor.
  MY REQUEST is for you to take me step by step
  to create a strategic plan that will improve
  overall company performance as the COO. YOUR
  RESPONSE will be to lead me one step at a time
  until we accomplish the following: (1) assess
  current operations, (2) gather insights from
  department heads, (3) identify key areas for
  improvement, and (4) draft a structured strategic
  plan with actionable steps to enhance operational
  efficiency and employee engagement.
```

- Step 4: Identify Key Areas for Improvement: Based on the data and feedback gathered, identify the key areas where improvements are needed. Focus on areas that will have the greatest impact on operational efficiency and employee engagement.
- Step 5: Draft the Strategic Plan: Develop a detailed strategic plan that includes specific, measurable goals, timelines, and responsibilities for each department. Ensure that the plan aligns with the company's overall strategic objectives and includes metrics for tracking progress.
- Step 6: Communicate the Plan to Key Stakeholders: Present the strategic plan to the company's leadership team and key stakeholders. Ensure that everyone understands their roles and responsibilities in executing the plan and how it aligns with the company's broader goals.
- Step 7: Implement and Monitor the Plan: Begin implementing the strategic plan, working closely with department heads to ensure that the plan is executed effectively. Regularly monitor progress and make adjustments as needed to stay on track with the company's performance goals.

- *Results*
  - Developed a comprehensive strategic plan that improved overall company performance by [X%] within the first year.
  - Enhanced operational efficiency across all departments by [Y%], resulting in savings of approx [$X.X million] annually.
  - Increased employee engagement scores by [Z%], with [XX%] of employees reporting improved alignment with company goals in the annual survey.

Case Study: CFO Evaluation of Company Financials

Situation

- Challenge: As the Chief Financial Officer (CFO), you are responsible for assessing the company's financial health and providing a strategic evaluation to senior leadership. You need to analyze key financial statements—including the income statement, balance sheet, and cash flow statement—to identify areas for cost savings, evaluate the company's financial stability, and forecast future performance. This evaluation will be critical for the upcoming board meeting, where strategic decisions will be made.
- Goals:
  - Evaluate the company's financial statements.
  - Identify cost-saving opportunities amounting to [X%] of the total operating expenses.
  - Provide a financial forecast for the next fiscal year with expected revenue growth of [Y%].

Tools

- Potential Technology: LLM (ChatGPT, Claude, Gemini, Llama, xAI), Financial Analysis Software (e.g., SAP, Oracle Finan-

Case Studies For Your Career

cials), Microsoft Excel, Forecasting Tools.
- Potential Team: CFO, Financial Analysts, Accounting Team, Department Heads (for input on cost-saving opportunities).

Action

- Step 1: Gather and Analyze Financial Statements: Begin by collecting the most recent income statement, balance sheet, and cash flow statement. Use financial analysis software and Excel to thoroughly evaluate the company's financial performance, focusing on revenue streams, cost structures, asset management, and cash flow.
- Step 2: Engage with Department Heads: Meet with department heads to gain insights into departmental budgets, expenditures, and potential areas for cost savings. Ensure that the financial data aligns with the company's operational strategies and goals.
- Step 3: Use Your LLM to Draft the Financial Evaluation: Use your LLM to help draft a comprehensive financial evaluation report. The First Prompt is provided below. Iterate as needed.

```
First Prompt Example:
> YOUR ROLE is an AI financial advisor. MY REQUEST
  is for you to guide me step by step in evaluating
  the company's financial health in my role as CFO.
  YOUR RESPONSE will begin by asking me to upload
  the necessary Excel files containing the core
  financial statements, such as income statements,
  balance sheets, and cash flow statements.
  Additionally, provide me with a comprehensive
  list of other relevant financial documents
  that could be important to evaluate, including
  supplementary documents like budget vs. actual
  reports, accounts receivable aging reports,
  inventory turnover reports, debt schedules,
```

> and capital expenditure (CapEx) reports. After gathering these documents, please guide me through reviewing and analyzing them to identify cost-saving opportunities and prepare a financial evaluation report.

- Step 4: Identify Cost-Saving Opportunities: Based on the financial analysis and input from department heads, identify key areas where the company can reduce costs by [X%]. Focus on optimizing operations, reducing waste, and improving financial efficiency without compromising quality or performance.
- Step 5: Draft the Financial Forecast: Develop a financial forecast for the next fiscal year, projecting revenue growth of [Y%] based on current trends, market conditions, and strategic initiatives. Ensure that the forecast is realistic and aligns with the company's overall strategic goals.
- Step 6: Prepare for the Board Meeting: Compile the financial evaluation, cost-saving recommendations, and forecast into a comprehensive report. Prepare to present this report to the board, emphasizing the strategic importance of the findings and recommendations.

Results

- Evaluated the company's financial statements, identifying key areas of financial strength and concern.
- Identified cost-saving opportunities amounting to [X%] of total operating expenses, leading to potential savings of [$X.X million] annually.
- Provided a financial forecast projecting [Y%] revenue growth for the next fiscal year, contributing to strategic decision-making at the board level.

Case Study: CTO Business Case for Integrating AI into Company Culture

Situation

- Challenge: As the Chief Technology Officer (CTO), you recognize the transformative potential of AI and want to integrate AI into the company's culture to drive innovation, improve operational efficiency, and maintain a competitive edge. You are tasked with creating a business case that outlines the benefits, costs, and strategic alignment of integrating AI into the company's operations and culture.
- Goals:
  - Develop a comprehensive business case for integrating AI into the company's culture.
  - Demonstrate potential increases in operational efficiency by [X%] and reductions in costs by [Y%].
  - Highlight how AI integration will enhance the company's competitive edge and drive innovation.

Tools

- Potential Technology: LLM (ChatGPT, Claude, Gemini, Llama, xAI), AI Strategy Tools, Data Analysis Tools (e.g., Excel, Tableau), Project Management Software (e.g., Asana, Jira).
- Potential Team: CTO, IT Department, AI Specialists, Department Heads, HR (for cultural integration).

Action

- Step 1: Assess Current State of AI Usage: Begin by evaluating the current use of AI within the company. Identify existing AI tools and technologies in use, as well as potential areas where

AI could be integrated to improve efficiency, innovation, and decision-making.
- Step 2: Engage with Key Stakeholders: Meet with department heads and key stakeholders to understand their needs, challenges, and opportunities for AI integration. Gather input on how AI can be leveraged to enhance workflows, drive innovation, and support strategic goals.
- Step 3: Use Your LLM to Draft the Business Case: Use your LLM to help draft a comprehensive business case for integrating AI into the company's culture. The First Prompt is provided below. Iterate as needed.

```
First Prompt Example:
> YOUR ROLE is an AI business strategist. MY
REQUEST is for you to take me step by step to
create a business case for integrating AI into
our company culture. YOUR RESPONSE will be to
lead me one step at a time until we accomplish
the following: (1) assess the current state of AI
usage, (2) gather input from key stakeholders,
(3) identify key areas where AI can drive
efficiency and innovation, and (4) draft a
structured business case that demonstrates the
strategic value of AI integration.
```

- Step 4: Identify Key Areas for AI Integration: Based on the assessment and stakeholder input, identify the key areas where AI can be integrated to drive the most significant improvements. Focus on areas such as operational efficiency, customer experience, product innovation, and data-driven decision-making.
- Step 5: Develop the Business Case: Draft a comprehensive business case that includes an analysis of the potential benefits, costs, and risks associated with AI integration. Highlight how

AI can improve operational efficiency by [X%], reduce costs by [Y%], and enhance the company's competitive edge.
- Step 6: Align AI Integration with Company Culture: Work with HR and department heads to ensure that AI integration aligns with the company's culture and values. Develop a plan for training, communication, and change management to ensure a smooth transition.
- Step 7: Present the Business Case to Leadership: Prepare to present the business case to the company's leadership team, emphasizing the strategic importance of AI integration and the potential for long-term value creation. Be prepared to address any concerns or questions about costs, risks, and cultural impact.

Results

- Developed a comprehensive business case for integrating AI into the company's culture, demonstrating the strategic value of AI adoption.
- Identified key areas where AI integration could improve operational efficiency by [X%] and reduce costs by [Y%], with potential savings of [$X.X million] annually.
- Presented the business case to leadership, securing buy-in for the strategic integration of AI into company culture, driving innovation, and maintaining a competitive edge.

Case Study: CEO Navigating a Major Merger or Acquisition

Situation

- Challenge: As the CEO, you are leading the company through a major merger or acquisition. This strategic move is intended to expand market share, acquire new technologies, or enter new markets. However, the success of this merger or acquisition depends on thorough due diligence, effective cultural integration, clear communication, and maximizing the synergies between the two organizations. You need to ensure that this complex process aligns with the company's long-term goals and creates maximum value for shareholders.

- Goals:
  - Successfully complete the merger or acquisition with thorough due diligence and strategic alignment.
  - Ensure smooth cultural integration between the merging companies, with [X%] employee retention.
  - Achieve expected synergies and cost savings amounting to [Y%] of the combined operating expenses.

Tools

- Potential Technology: LLM (ChatGPT, Claude, Gemini, Llama, xAI), M&A Analysis Tools, Financial Modeling Software, Cultural Assessment Tools.
- Potential Team: CEO, CFO, Legal Team, HR (for cultural integration), M&A Advisory Team, Communication Team.

Action

- Step 1: Conduct Thorough Due Diligence: Begin by conducting an exhaustive due diligence process, examining the target company's financial health, legal standing, operational capabilities, and cultural fit. Use M&A analysis tools and financial modeling software to assess the value and risks associated with the merger or acquisition.
- Step 2: Engage with Key Stakeholders: Communicate with key

Case Studies For Your Career 193

stakeholders, including the board of directors, senior leadership, and shareholders, to ensure alignment with the company's strategic goals. Gather input on potential risks, opportunities, and the desired outcomes of the merger or acquisition.
- Step 3: Use Your LLM to Draft the M&A Strategy: Use your LLM to help draft a comprehensive M&A strategy. The First Prompt is provided below. Iterate as needed.

```
First Prompt Example:
> YOUR ROLE is an AI M&A strategist. MY REQUEST
  is for you to take me step by step to navigate
  a major merger or acquisition as the CEO. YOUR
  RESPONSE will be to lead me one step at a time
  until we accomplish the following: (1) conduct
  thorough due diligence, (2) engage with key
  stakeholders, (3) develop a detailed integration
  plan, and (4) draft a structured M&A strategy
  that aligns with the company's long-term goals
  and maximizes shareholder value.
```

- Step 4: Develop a Detailed Integration Plan: Create an integration plan that addresses key areas such as cultural alignment, operational integration, IT systems, and employee retention. Focus on maintaining business continuity and minimizing disruptions during the transition period.
- Step 5: Communicate the Plan to All Stakeholders: Develop a clear communication strategy to inform all stakeholders—employees, customers, shareholders, and partners—about the merger or acquisition. Ensure transparency and provide regular updates on the progress and impact of the integration.
- Step 6: Monitor and Adjust the Integration Process: Oversee the execution of the integration plan, closely monitoring key

performance indicators (KPIs) related to synergies, cost savings, and employee retention. Be prepared to make adjustments as needed to address challenges and ensure the merger or acquisition achieves its strategic objectives.

Results

- Successfully completed the merger or acquisition, with due diligence ensuring minimal risks and strategic alignment.
- Achieved smooth cultural integration with [X%] employee retention, fostering a unified organizational culture.
- Realized expected synergies and cost savings amounting to [Y%] of the combined operating expenses, resulting in [$X.X million] in annual savings.

CHAPTER 12

CASE STUDIES FOR YOUR FAMILY

PARENTING, FAMILY LIFE, & HEALTH/WELFARE

The complexities of modern living often require innovative solutions to manage daily tasks, foster healthy relationships, and ensure the well-being of every family member. Artificial Intelligence offers a unique and powerful set of tools that can help families navigate these challenges, providing support in areas ranging from parenting and home management to health and wellness.

This chapter presents case studies showcasing AI integration into family life, using the STAR format—Situation, Tools, Action, and Result. These studies provide practical insights for parenting, family organization, home management, health and welfare.

Parenting

Using AI can help offer parents innovative ways to support their children's academic development, ensure responsible use of technology, and continue their own lifelong learning journey while balancing family responsibilities.

Case Study: Academic Struggling Student

Situation

- Challenge: Your child is struggling with their schoolwork, and you want to provide them with support without overwhelming them or yourself. You want to use AI and other tools to help them stay on top of assignments, improve their academic performance, and establish a long-term strategy for their academic career.
- Goals:
  - Improve your children's grades from # to # in the subjects of A, B, C, D.
  - Establish a consistent homework routine.
  - Reduce homework-related stress by 50%.
  - Foster independent learning skills in your children.

Tools

- Potential Technology: LLM (ChatGPT, Claude, Gemini, Llama, xAI), Online Video (e.g., Khan Academy, YouTube, Duolingo), Microsoft Office Suite
- Potential Team: Parents, children, tutors, teachers

Action

- Step 1: Identify the subjects and topics where your children need the most help.
- Step 2: Discuss with your child their current grades and goals for future progress. Put it in writing and reflect on it often.
- Step 3: Use your LLM to assist you and your child in crafting a comprehensive strategy. An example prompt is provided below. Iterate as needed.

```
First Prompt Exempt:
> Your Role is an AI homework assistant. My Request
  is for you to help my children with their
  homework by providing explanations, examples,
  and practice problems tailored to their needs.
  Your Response will be to (1) deliver a detailed
  time-bound and measurable plan for assisting with
  specific subjects and topics, (2) agreed grade
  progress goals.
```

- Step 4: Once the LLM plan is produced, set up a regular daily study schedule to cultivate effective study habits.
- Step 5: Offer positive feedback and modify your AI plan according to your child's progress.
- Step 6: Motivate the children to partake in self-directed learning by utilizing AI tools for responsible, independent exploration of new subjects.
- Step 7: Celebrate success!

Results
- Improved your children's grades by agreed goals in each subject.
- Established a consistent and effective homework routine.
- Your children say they are no longer stressed doing homework.
- Your children can complete homework without supervision.

Case Study: Responsible AI Use For Families

Situation

- Challenge: With the increasing use of AI in daily life, you want to ensure that your family is using AI tools responsibly and ethically. This involves educating family members about AI's capa-

bilities and limitations, establishing guidelines for AI use, and fostering a healthy relationship with technology.
- Goals:
  - Educate all family members about the capabilities, benefits, and risks of AI.
  - Establish clear guidelines for responsible and ethical AI use in the household.
  - Foster a healthy balance between AI use and offline activities.

Tools

- Potential Technology: LLM (ChatGPT, Claude, Gemini, Llama, xAI), Parental Control Apps, AI Monitoring Tools, Online Educational Resources (e.g., AI Ethics courses, YouTube), Family Agreement Templates.
- Potential Team: Parents, children, family members.

Action

- Step 1: Educate yourself about AI's capabilities, benefits, and risks using online resources. Then, share this knowledge with your family in an age-appropriate way.
- Step 2: Hold a family meeting to discuss the importance of responsible AI use. Outline the benefits and potential risks, and explain why guidelines are necessary.
- Step 3: Use your LLM to help draft a family AI use policy. An example prompt is provided below. Iterate as needed.

```
First Prompt Example:
> Your Role is an AI ethics advisor. My Request
  is for you to help me draft a family policy on
  responsible and ethical AI use. Your Response
  will be to (1) provide a detailed set of
```

Case Studies For Your Family

> guidelines covering privacy, data security, screen time, and ethical considerations, and (2) suggest strategies for monitoring and enforcing these guidelines.

- Step 4: Implement the AI use policy by integrating it into daily routines. Ensure that all family members understand and agree to the guidelines.
- Step 5: Encourage regular family discussions about AI use, including any concerns or questions that arise, and balance AI use with offline activities like outdoor play, family games, and reading.

Results

- Educated all family members about the capabilities, benefits, and risks of AI.
- Established and implemented clear guidelines for responsible and ethical AI use.
- Fostered a healthy balance between AI use and offline activities.

Case Study: Lifelong Learning for Parents

Situation

- Challenge: As a parent, you want to continue your personal and professional development while managing family responsibilities. You're looking for AI tools that can help you engage in continuous learning and skill development at your own pace, without disrupting your family life.
- Goals:
  - Identify key areas for personal and professional growth.

- Complete one new course or learning module per month.
- Integrate learning activities into your daily routine without disrupting family time.

Tools

- Potential Technology: LLM (ChatGPT, Claude, Gemini, Llama, xAI), AI-driven learning platforms (e.g., Coursera, LinkedIn Learning), Online Video (e.g., YouTube, Skillshare), Microsoft Office Suite.
- Potential Team: Individual (Parent), learning community or peers (optional).

Action

- Step 1: Assess your current skills and identify areas for improvement or new learning. Use your LLM to explore potential learning paths and identify courses that align with your goals.
- Step 2: Discuss with your family the importance of personal development and how you plan to integrate learning into your daily routine.
- Step 3: Use your LLM to assist in crafting a personalized learning plan. Iterate as needed.

```
First Prompt: YOUR ROLE is an AI learning advisor.
My Request is for you to help me create a
personalized learning plan that fits my schedule
and goals. Your Response will be to (1) provide
a list of relevant courses, learning paths, and
resources, and (2) suggest a monthly schedule
that balances learning with family time.
```

- Step 4: Implement the AI-generated learning plan, scheduling regular study sessions that align with your family's routine.

- Step 5: Use the AI tools to track progress, suggest additional resources, and adjust the learning plan as needed. Apply newly acquired skills in real-life situations, personally and professionally.

Results

- Identified key areas for personal and professional growth.
- Completed one new course or learning module per month.
- Successfully integrated learning activities into your daily routine without disrupting family time.

Family Organization

For many families, managing day-to-day life can feel like juggling a hundred balls in the air. The AI-enabled family has more than just a smooth schedule—they gain back hours in their week and dollars in their budget.

Case Study: Planning Family Vacations

Situation

- Challenge: Planning a family vacation can be stressful and time-consuming, with various factors to consider, such as budget, activities, and schedules. You want to use AI tools to organize and plan a stress-free family vacation that everyone will enjoy.
- Goals:
  - Plan a family vacation within budget.
  - Ensure the vacation meets the interests and needs of all family members.
  - Reduce vacation planning stress by 50%.

Tools

- Potential Technology: LLM (ChatGPT, Claude, Gemini, Llama, xAI), AI travel planning apps (e.g., Hopper, TripIt), budget management tools, AI itinerary planners, Microsoft Office Suite.
- Potential Team: Parents, children, travel agents (optional).

Action

- Step 1: Discuss with your family to identify vacation preferences, including budget, destinations, and activities. Use an LLM to gather ideas and preferences.
- Step 2: Set up AI travel planning apps and input your vacation preferences, including budget and desired activities.
- Step 3: Use your LLM to help craft a comprehensive travel plan. An example prompt is provided below. Iterate as needed.

```
First Prompt Example:
> YOUR ROLE is an AI travel planner. MY REQUEST
  is for you to help us plan a stress-free family
  vacation that fits our budget and meets the
  interests of all family members. YOUR RESPONSE
  will be to (1) provide a detailed travel
  plan, including destinations, activities, and
  logistics, and (2) suggest cost-saving options
  and family-friendly activities.
```

- Step 4: Implement the AI-generated travel plan by booking flights, accommodations, and activities that align with your budget and family's interests.
- Step 5: Develop a flexible itinerary that balances relaxation with planned activities, using AI itinerary planners to keep everything organized.

Results

- Planned a family vacation within budget.
- Ensured the vacation met the interests and needs of all family members.
- Reduced vacation planning stress by 50%.

Case Study: Financial Planning for Families

Situation

- Challenge: Managing family finances can be complex, especially when trying to balance budgeting, saving, and planning for future expenses. You want to leverage AI-driven tools to create a financial plan that meets your family's goals and ensures financial stability.
- Goals:
    - Develop a comprehensive family budget that aligns with your financial goals.
    - Increase monthly savings by 20%.
    - Ensure bills and payments are managed efficiently and on time.

Tools

- Potential Technology: LLM (ChatGPT, Claude, Gemini, Llama, xAI), AI-driven budgeting apps (e.g., YNAB, Mint), investment management tools, Microsoft Office Suite.
- Potential Team: Parents, financial advisor (optional).

Action

- Step 1: Assess your current financial situation, including income, expenses, savings, and debts. Use your LLM to help analyze and organize this data.

- Step 2: Discuss with your family your financial goals and priorities, ensuring everyone is aligned.
- Step 3: Use your LLM to help craft a comprehensive financial plan. An example prompt is provided below. Iterate as needed.

```
First Prompt Example:
> YOUR ROLE is an AI financial advisor. My Request
  is for you to help us create a comprehensive
  family budget and savings plan that aligns with
  our financial goals. Your Response will be to (1)
  analyze our financial data and provide a tailored
  budget and savings strategy, and (2) suggest ways
  to optimize expenses and increase savings.
```

- Step 4: Implement the AI-generated financial plan by setting up automated bill payments and savings contributions.
- Step 5: Schedule monthly reviews of your budget and financial plan, making adjustments as necessary to stay on track with your goals.

Results

- Developed a comprehensive family budget aligned with financial goals.
- Increased monthly savings by 20%.
- Ensured bills and payments were managed efficiently and on time.

Home Management

These case studies show how AI can effectively optimize home energy consumption, enhance security, and streamline home maintenance, resulting in a safer, more efficient, and well-maintained household.

Case Study: Home Energy Management

Situation

- Challenge: Your household energy consumption is higher than you'd like, leading to increased utility bills and a larger carbon footprint. You want to use AI tools to monitor and reduce energy consumption, making your home more energy-efficient and environmentally friendly.
- Goals:
  - Reduce household energy consumption by 25%.
  - Lower utility bills by 20%.
  - Identify and address energy inefficiencies in the home.

Tools

- Potential Technology: LLM (ChatGPT, Claude, Gemini, Llama, xAI), AI-powered energy management systems (e.g., Sense, Google Nest), smart thermostats, energy monitoring apps.
- Potential Team: Homeowner, energy consultant (optional).

Action

- Step 1: Assess your current energy consumption and identify areas where improvements can be made. Use your LLM to analyze patterns and suggest areas for optimization.
- Step 2: Set up AI-powered energy management systems and connect them to your home's energy sources to monitor usage in real-time.
- Step 3: Use your LLM to help craft a comprehensive energy-saving strategy. Iterate as needed.

```
First Prompt Example:
> YOUR ROLE is an AI energy manager. My Request is
  for you to help us reduce our household energy
  consumption by monitoring usage, identifying
  inefficiencies, and providing recommendations
  for energy-saving actions. Your Response will
  be to (1) analyze our energy data and deliver a
  tailored energy management plan, and (2) suggest
  specific actions to reduce consumption and lower
  costs.
```

- Step 4: Implement the AI-generated energy management plan, such as optimizing heating/cooling schedules or reducing standby power consumption.
- Step 5: Review energy consumption data weekly to track progress and make further adjustments as to achieve your goals.

Results

- Reduced household energy consumption by 25%.
- Lowered utility bills by 20%.
- Identified and addressed energy inefficiencies in the home.

Case Study: Home Security Best Practices

Situation

- Challenge: Ensuring the safety and security of your home is a top priority, but traditional security systems can be limited in their responsiveness and adaptability. You want to enhance your home security with AI-powered tools that offer real-time monitoring, advanced threat detection, and automated responses.

Case Studies For Your Family

- Goals:
  - Increase home security and reduce the risk of break-ins by 50%.
  - Implement real-time monitoring and threat detection.
  - Automate security responses to potential threats.

Tools

- Potential Technology: LLM (ChatGPT, Claude, Gemini, Llama, xAI), AI-powered home security systems (e.g., Ring, SimpliSafe), smart cameras, motion sensors.
- Potential Team: Homeowner, security consultant (optional).

Action

- Step 1: Assess your current home security setup and identify areas for improvement. Use your LLM to analyze vulnerabilities.
- - Step 2: Set up AI-powered home security systems, including smart cameras and motion sensors to monitor your home.
- Step 3: Use your LLM to help craft a comprehensive security plan. An example prompt is provided below. Iterate as needed.

```
First Prompt:
> YOUR ROLE is an AI security advisor. My Request
  is for you to help us enhance our home security
  by providing real-time monitoring, advanced
  threat detection, and automated responses. Your
  Response will be to (1) design a comprehensive
  security plan tailored to our home, and (2)
  suggest specific measures to improve security and
  automate threat responses.
```

- Step 4: Implement the AI-generated security plan by integrating the security system with other smart home devices for enhanced automation.

- Step 5: Monitor real-time alerts and review security footage regularly to ensure the system is functioning as intended and to update the plan as necessary.

Results

- Increased home security and reduced the risk of break-ins by 50%.
- Implemented real-time monitoring and threat detection.
- Automated security responses to potential threats.

Case Study: Home Maintenance Scheduling

Situation

- Challenge: Keeping up with home maintenance tasks can be challenging, leading to deferred maintenance and unexpected repair costs. You want to use AI tools to automate the scheduling of home maintenance tasks, ensuring that your home remains in an optimal condition without the stress of manual tracking.
- Goals:
    - Ensure timely completion of all home maintenance tasks.
    - Reduce the likelihood of costly repairs by 30%.
    - Automate scheduling and reminders for regular maintenance.

Tools

- Potential Technology: LLM (ChatGPT, Claude, Gemini, Llama, xAI), AI-driven home maintenance apps (e.g., HomeZada, Centriq), smart home devices.
- Potential Team: Homeowner, maintenance professionals

Case Studies For Your Family

Action

- Step 1: Assess your home's maintenance needs, including routine tasks and potential issues. Use your LLM to help identify and prioritize these tasks.
- Step 2: Set up AI-driven home maintenance apps and input all relevant maintenance tasks and schedules.
- Step 3: Use your LLM to help craft a comprehensive maintenance plan. Iterate as needed.

```
First Prompt Example:
> YOUR ROLE is an AI home maintenance manager.
  My Request is for you to help us automate and
  schedule home maintenance tasks to ensure
  timely completion and prevent costly repairs.
  Your Response will be to (1) provide a tailored
  maintenance plan with automated scheduling and
  reminders, and (2) suggest specific actions for
  maintaining home systems and appliances.
```

- Step 4: Implement the AI-generated maintenance plan, ensuring all tasks are automated and tracked through the app.
- Step 5: Conduct regular reviews of maintenance logs and AI system performance to ensure all tasks are being completed as scheduled and that the system is functioning optimally.

Results

- Ensured timely completion of all home maintenance tasks.
- Reduced the likelihood of costly repairs by 30%.
- Automated scheduling and reminders for regular maintenance.

Case Study: Streamlining Household Chores

Situation

- Challenge: Managing household chores can be overwhelming, leading to inefficiencies and stress within the family. You want to utilize AI tools to organize and streamline these tasks, ensuring that chores are distributed fairly and completed efficiently.
- Goals:
  - Reduce time spent on household chores by 30%.
  - Ensure fair distribution of chores among family members.
  - Increase chore completion rate to 95%.

Tools

- Potential Technology: LLM (ChatGPT, Claude, Gemini, Llama, xAI), AI-powered home management apps (e.g., Tody, OurHome), task automation tools, smart home devices.
- Potential Team: Family members.

Action

- Step 1: Assess the current distribution and efficiency of household chores. Use your LLM to help identify areas for improvement and suggest ways to optimize the process.
- Step 2: Discuss with your family the importance of sharing household responsibilities and set goals together for improving efficiency and fairness.
- Step 3: Use your LLM to help craft a comprehensive chore management strategy. An example prompt is provided below. Iterate as needed.

```
First Prompt Example:
> Your Role is an AI household manager. My Request
  is for you to help us streamline and automate
  household chores to ensure fair distribution
  and efficient completion. Your Response will
  be to (1) create a tailored chore schedule with
  automation options where applicable, and (2)
  suggest tools or devices that can further reduce
  manual effort.
```

- Step 4: Set up AI-powered home management apps to assign and track household chores. Integrate smart home devices to automate repetitive tasks, such as vacuuming or laundry.
- Step 5: Monitor chore completion through the AI app and adjust the schedule as needed to maintain efficiency and fairness.

Results

- Reduced time spent on household chores by 30%.
- Ensured fair distribution of chores among family members.
- Increased chore completion rate to 95%.

Health and Wellness

These case studies under the Health and Wellness category show how AI can effectively monitor family health, encourage healthy habits, and support elderly family members, resulting in a healthier and more harmonious family environment.

Case Study: Monitoring Family Health

Situation

- Challenge: Keeping track of the health metrics of all family members can be challenging, especially when trying to manage multiple schedules and health concerns. You want to use AI-powered health trackers to monitor physical health indicators such as heart rate, sleep patterns, and activity levels, helping your family stay on top of their health.

- Goals:
  - Track key health metrics (e.g., heart rate, sleep, activity) for all family members.
  - Identify and address any concerning health trends early on.
  - Increase daily physical activity for each family member by 20%.

Tools

- Potential Technology: LLM (ChatGPT, Claude, Gemini, Llama, xAI), AI-powered health trackers (e.g., Fitbit, Apple Watch), health monitoring apps.
- Potential Team: Family members, healthcare provider

Action

- Step 1: Assess the current health status and needs of each family member, including any specific health concerns. Use your LLM to analyze and organize this data.
- Step 2: Set up AI-powered health trackers for each family member, ensuring they are customized to monitor relevant health metrics.
- Step 3: Use your LLM to help craft a comprehensive health monitoring strategy. An example prompt is provided below. Iterate as needed.

```
First Prompt Example:
> Your Role is an AI health advisor. My Request
  is for you to help us monitor and improve the
```

> health of our family by tracking key metrics and providing actionable insights. Your Response will be to (1) design a personalized health monitoring plan tailored to each family member, and (2) suggest daily activities to increase physical activity and improve overall health.

- Step 4: Implement the AI-generated health monitoring plan, tracking and reviewing health metrics daily through the health monitoring apps.
- Step 5: Schedule regular family check-ins to review health data, discuss any necessary changes, and celebrate progress in meeting health goals.

Results

- Successfully tracked key health metrics for all family members.
- Identified and addressed concerning health trends early.
- Increased daily physical activity for each family member by 20%.

Case Study: Encouraging Healthy Eating Habits

Situation

- Challenge: Ensuring that your family maintains healthy eating habits can be difficult, especially with busy schedules and differing dietary preferences. You want to use AI tools to plan nutritious meals, track food intake, and encourage healthy eating across the family.
- Goals:
    - Plan balanced, nutritious meals for the entire family.
    - Increase the consumption of fruits and vegetables by 30%.
    - Reduce the intake of processed foods and sugary snacks by 40%.

Tools

- Potential Technology: LLM (ChatGPT, Claude, Gemini, Llama, xAI), AI meal planning apps (e.g., Eat This Much, Yummly), nutrition tracking apps.
- Potential Team: Family members, nutritionist (optional).

Action

- Step 1: Assess the current dietary habits of each family member and identify areas for improvement. Use your LLM to help analyze eating patterns and suggest healthier alternatives.
- Step 2: Set up AI meal planning and nutrition tracking apps to create personalized meal plans and monitor food intake.
- Step 3: Use your LLM to help craft a comprehensive meal planning and nutrition strategy. An example prompt is provided below. Iterate as needed.

```
First Prompt Example:
> Your Role is an AI nutrition advisor. My Request
  is for you to help us plan healthy, balanced
  meals and track our food intake to improve
  overall family nutrition. Your Response will be
  to (1) provide meal plans, recipes, and tracking
  tools tailored to our dietary needs, and (2)
  suggest strategies to increase the consumption of
  healthy foods while reducing unhealthy ones.
```

- Step 4: Implement the AI-generated meal plans, using the apps to monitor daily food intake and provide insights into nutritional balance.
- Step 5: Hold weekly family meetings to review progress, discuss challenges, and adjust meal plans as needed to meet health goals.

Results

- Planned balanced, nutritious meals for the entire family.
- Increased the consumption of fruits and vegetables by 30%.
- Reduced the intake of processed foods and sugary snacks by 40%.

Case Study: Supporting Elderly Family Members

Situation

- Challenge: As your elderly family members age, ensuring they stay connected, engaged, and well-cared for can be challenging. You want to use AI tools to monitor their health, assist with daily tasks, and help them maintain social connections, improving their quality of life.
- Goals:
  - Monitor the health and well-being of elderly family members.
  - Increase engagement in social and cognitive activities by 50%.
  - Automate reminders for medications, appointments, and daily tasks.

Tools

- Potential Technology: LLM (ChatGPT, Claude, Gemini, Llama, xAI), AI-powered health monitoring devices, social engagement apps, virtual assistants (e.g., Amazon Alexa, Google Assistant).
- Potential Team: Family members, caregivers, healthcare provider (optional).

Action

- Step 1: Assess the current health, social engagement, and daily routine needs of your elderly family members. Use your LLM to help organize and prioritize these needs.

- Step 2: Set up AI-powered health monitoring devices to track vital signs, mobility, and daily activities.
- Step 3: Use your LLM to help craft a comprehensive care plan. An example prompt is provided below. Iterate as needed.

```
First Prompt Example:
> Your Role is an AI elder care assistant. My
  Request is for you to help us monitor and support
  our elderly family members, ensuring they remain
  healthy, engaged, and connected. Your Response
  will be to (1) provide a comprehensive care plan
  with monitoring, reminders, and social engagement
  activities, and (2) suggest ways to enhance their
  quality of life through personalized care.
```

- Step 4: Implement the AI-generated care plan, including the use of virtual assistants to automate reminders for medications, appointments, and daily tasks.
- Step 5: Schedule regular check-ins with family members or caregivers to review their health and well-being, and adjust the care plan as needed.

Results

- Monitored the health and well-being of elderly family members.
- Increased engagement in social and cognitive activities by 50%.
- Successfully automated reminders for medications, appointments, and daily tasks.

Chapter 13

Case Studies for Your Faith

Prayer, Study, & Sharing Faith

Faith is an integral part of many people's lives, providing guidance, comfort, and a sense of community. However, maintaining and nurturing one's faith amidst the demands of modern life can be challenging. This chapter explores how artificial intelligence (AI) can support and enhance various aspects of your faith life, offering innovative solutions to age-old challenges.

Our case studies use the STAR format—Situation, Team/Tools, Action, and Result—to structure AI integration into your faith practices. Each starts with a detailed scenario outlining the CHALLENGE, followed by a description of the possible TOOLS. The ACTION section provides a step-by-step implementation plan, including an effective first prompt for AI use. Finally, we give you example RESULTS to measure success.

Whether you are someone wanting to strengthen your personal faith, a religious leader aiming to improve your ministry, or a community leader looking to build stronger bonds, these real-life examples provide practical and useful insights. By integrating AI into your faith life, you can overcome obstacles, achieve your spiritual goals, and create a more meaningful experience for yourself and those around you.

Prayer Case Studies

Prayer, a cornerstone of spiritual life, connects individuals with the divine, offers guidance, and brings peace. Modern life demands often hinder maintaining a consistent prayer routine and deepening spiritual practice. The following case studies show how AI can enhance prayer life by creating personalized devotionals, setting prayer reminders, and aiding in scripture memorization. Using AI tools fosters a more disciplined, focused, and meaningful prayer experience, strengthening one's faith.

Case Study: Daily Devotionals

Situation

- Challenge: You want to cultivate a consistent spiritual routine by incorporating daily devotionals into your life. However, with a busy schedule, it's challenging to find time, choose the right devotional content, and stay committed to the practice. You're looking for a way to integrate AI tools into your daily routine to help select meaningful devotional materials, remind you to engage in devotionals, and reflect on the spiritual lessons you learn.
- Goals:
  - Establish a consistent daily devotional practice, with [X] minutes dedicated each day.
  - Select devotional content that resonates with your personal spiritual journey.
  - Reflect on and apply spiritual lessons to your daily life.

Tools

- Potential Technology: LLM (ChatGPT, Claude, Gemini, Llama, xAI), Devotional Apps (e.g., YouVersion, Glorify), Calendar and Reminder Apps.

Case Studies For Your Faith

- Potential Team: Individual (Personal Spiritual Practice).

Action

- Step 1: Assess Your Spiritual Needs: Start by reflecting on your current spiritual journey and identifying areas where you seek growth or inspiration. Use your LLM to help suggest devotional themes or topics that align with your needs.
- Step 2: Select Devotional Content: Use devotional apps or your LLM to curate daily devotional content that resonates with you. This could include scripture readings, reflections, prayers, or inspirational stories.
- Step 3: Use Your LLM to Establish a Routine: Use your LLM to help create a daily devotional routine that fits your schedule. The First Prompt is provided below. Iterate as needed.

```
First Prompt Example:
> YOUR ROLE is an AI spiritual guide. MY REQUEST is
  for you to help me establish a consistent daily
  devotional practice. To begin, ask me reflective
  questions to assess my spiritual needs and
  goals. Use my responses to guide the selection
  of appropriate devotional content and to create
  a daily routine that aligns with my schedule
  and spiritual aspirations. After each response,
  provide guidance and ask follow-up questions if
  needed until we clearly understand my spiritual
  needs. Once that's complete, move on to selecting
  devotional content and creating a routine.
```

- Step 4: Set Daily Reminders: Schedule daily reminders through calendar apps or devotional apps to ensure that you set aside dedicated time for your devotionals each day.

- Step 5: Reflect and Apply: After each devotional, take a few moments to reflect on the spiritual lesson and how it can be applied to your daily life. Consider journaling your thoughts or discussing them with a spiritual mentor or community.

Results

- Established a consistent daily devotional practice, dedicating [X] minutes each day.
- Selected devotional content that resonates with your personal spiritual journey, enhancing the relevance and impact of your devotionals.
- Regularly reflected on and applied spiritual lessons to your daily life, deepening your spiritual growth.

Case Study: Daily Devotionals

Situation

- Challenge: You want to cultivate a consistent spiritual routine by incorporating daily devotionals into your life. However, with a busy schedule, it can be challenging to find time, choose the right devotional content, and stay committed to the practice. You're looking for a way to integrate AI tools into your daily routine to help select meaningful devotional materials, remind you to engage in devotionals, and reflect on the spiritual lessons.
- Goals:
  - Establish a consistent daily devotional practice, with [X] minutes dedicated each day.
  - Select devotional content that resonates with your personal spiritual journey.
  - Reflect on and apply spiritual lessons to your daily life.

Case Studies For Your Faith

Tools

- Potential Technology: LLM (ChatGPT, Claude, Gemini, Llama, xAI), Devotional Apps (e.g., YouVersion, Glorify), Calendar and Reminder Apps.
- Potential Team: Individual (Personal Spiritual Practice).

Action

- Step 1: Assess Your Spiritual Needs: Start by reflecting on your current spiritual journey and identifying areas where you seek growth or inspiration. Use your LLM to help suggest devotional themes or topics that align with your needs.
- Step 2: Select Devotional Content: Use devotional apps or your LLM to curate daily devotional content that resonates with you. This could include scripture readings, reflections, prayers, or inspirational stories.
- Step 3: Use Your LLM to Establish a Routine: Use your LLM to assist in creating a daily devotional routine that fits your schedule. The First Prompt is provided below. Iterate as needed.

```
First Prompt Example:
> YOUR ROLE is an AI spiritual guide. MY REQUEST
  is for you to take me step by step to establish
  a consistent daily devotional practice. YOUR
  RESPONSE will be to lead me one step at a time
  until we accomplish the following: (1) assess
  my spiritual needs, (2) select appropriate
  devotional content, and (3) create a daily
  routine that fits my schedule and goals.
```

- Step 4: Set Daily Reminders: Schedule daily reminders through calendar apps or devotional apps to ensure that you set aside dedicated time for your devotionals each day.

- Step 5: Reflect and Apply: After each devotional, take a few moments to reflect on the spiritual lesson and how it can be applied to your daily life. Consider journaling your thoughts or discussing them with a spiritual mentor or community.

Results

- Established a consistent daily devotional practice, dedicating [X] minutes each day.
- Selected devotional content that resonates with your personal spiritual journey, enhancing the relevance and impact of your devotionals.
- Regularly reflected on and applied spiritual lessons to your daily life, deepening your spiritual growth.

Case Study: Prayer Reminders

Situation

- Challenge: You want to strengthen your prayer life by ensuring that you pray regularly throughout the day. However, with a hectic schedule, it's easy to forget or push prayer aside. You need a system that integrates prayer reminders into your daily routine, helping you stay consistent and focused in your prayer practice.
- Goals:
  - Establish regular prayer times throughout the day, with [X] prayer sessions.
  - Create a habit of pausing for prayer during key moments of your daily routine.
  - Enhance the quality and depth of your prayer life.

Case Studies For Your Faith 223

Tools

- Potential Technology: LLM (ChatGPT, Claude, Gemini, Llama, xAI), Prayer Apps (e.g., PrayerMate, Echo), Calendar and Reminder Apps.
- Potential Team: Individual (Personal Spiritual Practice).

Action

- Step 1: Identify Key Moments for Prayer: Reflect on your daily routine and identify key moments where you can pause for prayer (e.g., morning, midday, evening). Use your LLM to help suggest optimal prayer times based on your schedule.

```
First Prompt Example:
> YOUR ROLE is an AI spiritual guide. MY REQUEST is
  for you to take me step by step to establish a
  regular prayer routine with reminders throughout
  the day. YOUR RESPONSE will be to lead me one
  step at a time until we accomplish the following:
  (1) identify key moments for prayer, (2) set up
  effective prayer reminders, and (3) enhance my
  prayer practice with personalized prayers or
  prompts.
```

- Step 2: Set Up Prayer Reminders: Use prayer apps or calendar apps to schedule reminders for each prayer session. Ensure that these reminders are integrated into your routine in a way that minimizes disruption.
- Step 3: Commit to the Practice: When each prayer reminder goes off, take a moment to pause, center yourself, and engage in prayer. Use this time to reconnect with your spiritual focus and intentions.

- Step 4: Reflect on Your Progress: Periodically reflect on your prayer practice to ensure it is meeting your spiritual needs. Adjust the timing, content, or focus of your prayers as needed to deepen your connection.

Results

- Established regular prayer times throughout the day, with [X] consistent prayer sessions.
- Created a habit of pausing for prayer during key moments of your daily routine, enhancing your spiritual connection.
- Enhanced the quality and depth of your prayer life through personalized prayers and focused spiritual practice.

Case Study: Scripture Memorization

Situation

- Challenge: You want to deepen your understanding and connection with scripture by committing key verses to memory. However, memorization can be challenging, especially with a busy schedule. You need an effective system that helps you select, memorize, and recall scripture verses as part of your growth.
- Goals:
    - Memorize [X] key scripture verses over a set period.
    - Improve recall and understanding of scripture through consistent practice.
    - Integrate scripture memorization into your daily routine.

Tools

- - Potential Technology: LLM, Scripture Memorization Apps (e.g., Bible Memory, Verses), Flashcard Apps (e.g., Anki).
- - Potential Team: Individual (Personal Spiritual Practice).

Case Studies For Your Faith

Action

- Step 1: Select Key Verses for Memorization: Begin by selecting a set of scripture verses that you want to commit to memory. Use your LLM to help suggest verses that are meaningful to you or aligned with your current spiritual focus.

```
First Prompt Example:
> YOUR ROLE is an AI memory coach. MY REQUEST is
  for you to take me step by step to memorize key
  scripture verses effectively. YOUR RESPONSE
  will be to lead me one step at a time until we
  accomplish the following: (1) select meaningful
  scripture verses, (2) create a structured
  memorization plan, and (3) enhance memorization
  with mnemonic devices or visualizations.
```

- Step 2: Practice Daily: Integrate scripture memorization into your daily routine, using apps or flashcards to review and reinforce the verses regularly. Consistency is key to committing the verses to long-term memory.
- Step 3: Reflect and Apply: As you memorize each verse, take time to reflect on its meaning and how it applies to your life. Consider discussing the verses with a study group or mentor to deepen your understanding.

Results

- Memorized [X] key scripture verses over the set period.
- Improved recall and understanding of scripture through consistent daily practice.
- Successfully integrated scripture memorization into your daily routine, enhancing your spiritual growth.

Bible Study

Engaging with scripture is vital for spiritual growth and communal faith. However, organizing Bible study, preparing sermons, and developing educational programs can be challenging. The following case studies show how AI can enhance Bible study, streamline sermon preparation, and create interactive education. Integrating AI fosters a more insightful and enriching experience for everyone.

Case Study: Small Group Bible Study Facilitator

Situation

- Challenge: As a small group leader, you are responsible for preparing and facilitating Bible study sessions that are engaging, insightful, and spiritually enriching for your group members. However, the preparation process can be time-consuming, especially for selecting discussion questions, providing relevant background information, and ensuring that the study remains focused and impactful. You want to use AI tools to assist in the preparation and facilitation of your small group Bible studies, helping you to create discussions and track progress.
- Goals:
  - Use AI to prepare discussion questions and provide background information on Bible passages.
  - Facilitate engaging and insightful Bible study sessions.
  - Track the group's progress and ensure that each study session builds on the previous ones.

Tools

- Potential Technology: LLM (ChatGPT, Claude, Gemini, Llama, xAI), Bible Study Software (e.g., Logos, Accordance).
- Potential Team: Small group leader, Group members.

Action

- Step 1: Identify the Study Focus and Passages: Begin by identifying the Bible passages or themes that your group will study. Use your LLM to help select passages that apply to the group's spiritual journey and to generate a set of discussion questions that will guide the study session. The questions should encourage deep reflection, group interaction, and application of the scripture to everyday life.

```
First Prompt Example:
> YOUR ROLE is an AI Bible study facilitator. MY
  REQUEST is for you to help me prepare for a small
  group Bible study session. YOUR RESPONSE will be
  to lead me one step at a time until we accomplish
  the following: (1) identify the study focus
  and relevant passages, (2) generate discussion
  questions that encourage deep reflection
  and interaction, and (3) provide background
  information that enhances the understanding of
  the passages.
```

- Step 2: Facilitate the Bible Study Session: During the Bible study session, use the AI-generated discussion questions and background information to guide the conversation. Encourage group members to share their insights and reflections, ensuring that everyone takes part.
- Step 3: Track Group Progress: After the session, use progress tracking tools to record the important points, discussion points, and any prayer requests or follow-up actions. This will help you monitor the group's spiritual growth over time and ensure that each session builds on the previous ones.

- Step 4: Adjust and Iterate: Based on the feedback and progress tracking, adjust future Bible study sessions to better meet the needs of the group. Use AI tools to continually refine your approach and enhance the overall study experience.

Results

- Prepared and facilitated Bible study sessions that were engaging, insightful, and spiritually enriching for the group.
- Used AI-generated discussion questions and background information to create meaningful discussions and deepen the group's understanding of the Bible passages.
- Successfully tracked the group's progress, ensuring that each study session built on the previous ones and contributed to the group's spiritual growth.

Case Study: AI-Driven Group Growth and Recruitment

Situation

- Challenge: As a small group leader, you want to grow your group by attracting new members who align with the group's dynamics and spiritual focus. However, identifying potential new members and reaching out to them in a targeted way can be challenging. You want to use AI tools to identify individuals who might benefit from joining your group based on their interests, spiritual needs, and demographic information. The AI can assist in creating targeted invitations and outreach efforts to ensure new members integrate well with the group.
- Goals:
  - Use AI to identify potential new members for the small group based on relevant criteria.

- Increase group membership by [X%] while maintaining group dynamics and cohesion.
- Enhance the group's spiritual growth and diversity by incorporating new members.

Tools

- Potential Technology: LLM (ChatGPT, Claude, Gemini, Llama, xAI), Church CRM Systems, Outreach and Communication Tools.
- Potential Team: Small group leader, Church leadership, Outreach team.

Action

- Step 1: Identify Group Dynamics and Needs: Start by assessing your current group's dynamics, including its spiritual focus, strengths, and areas where new members could contribute. Use AI tools to help analyze the group's characteristics and identify potential areas for growth.

> First Prompt Example:
> > YOUR ROLE is an AI group growth strategist. MY REQUEST is for you to help me grow my small group by identifying and recruiting potential new members. YOUR RESPONSE will be to lead me one step at a time until we accomplish the following: (1) assess the current group dynamics and needs, (2) identify potential new members based on relevant criteria, and (3) create targeted outreach to invite and integrate new members.

- Step 2: Create Targeted Outreach Efforts: Develop personalized invitations and outreach campaigns to connect with potential

new members. Use AI tools to tailor messages that resonate with the identified candidates, highlighting how the group can meet their spiritual needs.
- Step 3: Facilitate New Member Integration: Once new members join, focus on integrating them smoothly into the group. Use AI to suggest icebreaker activities, discussion topics, and group projects that help new members feel welcomed
- Step 4: Monitor Group Growth and Dynamics: Regularly assess the group's growth and dynamics after integrating new members. Use AI tools to gather feedback and make adjustments as needed to maintain group cohesion and spiritual growth.

Results

- Successfully identified and recruited new members, increasing group membership by [X%].
- Maintained group dynamics and cohesion while incorporating new members, enhancing the group's spiritual growth.
- Established a sustainable approach to group growth and recruitment, ensuring the ongoing vitality of the small group.

Sharing Faith

Sharing and living out faith within a community is key to building connections, fostering understanding, and making a positive impact. Whether organizing faith-based events, taking part in charity, or facilitating interfaith dialogue, these activities require careful planning. The case studies in this category illustrate how AI can assist in effectively sharing faith, from planning events to engaging in interfaith discussions and maximizing charitable impact. AI enhances the ability to live out faith and make a lasting community difference.

Case Studies For Your Faith

Case Study: AI-Enhanced Worship Experience

Situation

- Challenge: As a church leader, you aim to create worship services that deeply resonate with your congregation and foster a more meaningful spiritual experience. However, gathering and analyzing feedback, selecting the right songs, and structuring the service to meet the diverse needs of your congregation can be challenging. You want to use AI tools to enhance the worship experience by tailoring services based on congregational engagement and feedback data.
- Goals:
  - Use AI to analyze congregational feedback and engagement.
  - Tailor worship services, including the most compelling song selection, sermon topics, and service structures.
  - Increase overall congregational satisfaction and spiritual engagement by [X%].

Tools

- Potential Technology: LLM (ChatGPT, Claude, Gemini, Llama, xAI), Worship Planning Software, Feedback and Engagement Analysis Tools, Survey Platforms.
- Potential Team: Worship leader, Pastor, Tech team, Congregation feedback group.

Action

- Step 1: Gather Congregational Feedback and Engagement Data: Use survey platforms and engagement analysis tools to collect feedback from the congregation regarding worship services. Focus on elements like song selection, sermon impact, and overall service structure.

- Step 2: Analyze Data with AI: Use AI tools to analyze the collected data, identifying trends and preferences within the congregation. The AI can highlight the most impactful elements of past services and suggest areas for improvement. The First Prompt Example is provided below. Iterate as needed.

```
First Prompt Example:
> YOUR ROLE is an AI worship experience consultant.
  MY REQUEST is for you to help enhance our
  worship services by analyzing congregational
  feedback and engagement data. YOUR RESPONSE
  will be to lead me one step at a time until
  we accomplish the following: (1) identify key
  trends from congregational feedback, (2) suggest
  song selections, sermon topics, and service
  structures, and (3) tailor worship services to
  resonate more deeply with our congregation.
```

- Step 3: Implement Tailored Worship Services: Incorporate the AI-generated recommendations into your worship planning. Select songs, sermon topics, and service structures that align with the congregation's preferences and spiritual needs.
- Step 4: Monitor and Adjust: After implementing the tailored services, continue to gather feedback and monitor engagement. Use AI tools to adjust and refine future services based on ongoing data analysis.

Results

- Enhanced worship services by tailoring song selection, sermon topics, and service structures to better resonate.
- Increased congregational satisfaction and spiritual engagement by [X%].

- Established a continuous feedback loop to maintain and improve the quality of worship experiences.

Case Study: AI-Driven Sermon Content Creation

Situation

- Challenge: As a pastor, you strive to deliver sermons that are both spiritually enriching and relevant to your congregation. However, sermon preparation can be time-consuming, especially when it involves selecting scripture passages, researching historical context, and finding illustrative stories or examples. You want to use AI to assist in creating well-rounded, impactful sermons that connect with your congregation on a deeper level.
- Goals:
  - Streamline the sermon preparation process with AI assistance.
  - Enhance sermon content by incorporating relevant scripture, historical context, and illustrative stories.
  - Deliver sermons that resonate with the congregation and support spiritual growth by [X%].

Tools

- Potential Technology: LLM (ChatGPT, Claude, Gemini, Llama, xAI), Bible Study Software, Historical Context Databases, Illustration Libraries.
- Potential Team: Pastor, Worship leader, Tech team.

Action

- Step 1: Identify Sermon Themes and Scripture Passages: Begin by identifying the central theme or message of the sermon.
- Step 2: Use Your LLM to Create Sermon Content: Use your

LLM to help craft the sermon, incorporating the scripture, historical context, and illustrative stories. The First Prompt is provided below. Iterate as needed.

```
First Prompt Example:
> YOUR ROLE is an AI sermon content creator. MY
  REQUEST is for you to help me prepare a sermon by
  suggesting relevant scripture passages, providing
  historical context, and offering illustrative
  stories. YOUR RESPONSE will be to lead me one
  step at a time until we accomplish the following:
  (1) identify the central theme and related
  scripture, (2) research historical context, and
  (3) craft a well-rounded sermon that connects
  with the congregation.
```

- Step 3 Refine and Finalize the Sermon: Review the AI-generated content, making adjustments as needed to ensure the sermon aligns with your voice and the needs of your congregation.
- Step 4: Deliver and Gather Feedback: Deliver the sermon and gather feedback from the congregation. Use this feedback for future sermon preparation, continually refining your approach.

Results

- Streamlined the sermon preparation process, reducing time spent on research and content creation.
- Enhanced sermon content with well-selected scripture, historical context, and illustrative stories, leading to deeper congregation engagement.
- Increased the resonance and impact of sermons by [X%], supporting spiritual growth within the congregation.

Case Study: Congregational Engagement and Communication

Situation

- Challenge: Effective communication and engagement with your congregation are crucial for fostering a strong church community. However, with diverse member interests and varying levels of involvement, it's challenging to personalize messages, recommend relevant activities, and manage outreach efforts. You want to implement AI tools to improve communication strategies and enhance engagement within your church community.
- Goals:
  - Use AI to personalize communication and recommendations for church activities.
  - Increase overall congregational engagement by [X%].
  - Strengthen the church community through improved outreach and targeted messaging.

Tools

- Potential Technology: LLM (ChatGPT, Claude, Gemini, Llama, xAI), CRM Systems (e.g., Church Community Builder, Breeze), Social Media Management Tools.
- Potential Team: Church communications team, Pastoral staff, Tech team.

Action

- Step 1: Analyze Congregational Data: Use your church's CRM system to analyze congregational data, including attendance records, volunteer participation, and event involvement. Use data to segment congregation based on interests and engagement.

- Step 2: Use AI to Personalize Communication: Use your LLM to help create personalized communication strategies for different segments of the congregation. The AI can suggest relevant church activities, target messaging, and manage social media outreach to ensure each member feels connected and engaged.

```
First Prompt Example:
> YOUR ROLE is an AI communication strategist. MY
  REQUEST is for you to help improve congregational
  engagement by personalizing communication and
  recommending relevant church activities. YOUR
  RESPONSE will be to lead me one step at a time
  until we accomplish the following: (1) analyze
  congregational data to identify engagement
  patterns, (2) create personalized communication
  strategies, and (3) enhance outreach through
  targeted messaging.
```

- Step 3: Implement Personalized Communication Strategies: Begin sending personalized messages to different segments of the congregation, recommending activities and providing updates that are most relevant to their interests and involvement.
- Step 4: Manage Social Media Outreach: Use AI tools to optimize your church's social media presence, ensuring consistent and engaging content that fosters community interaction.
- Step 5: Monitor Engagement and Adjust: Regularly monitor the effectiveness of your communication strategies using engagement metrics. Adjust your approach based on feedback and data insights to continually improve engagement.

Results

- Increased overall congregational engagement by [X%] through personalized communication and targeted outreach.

- Strengthened the church community by connecting members with relevant activities and fostering deeper relationships.
- Enhanced the church's social media presence, leading to greater community interaction and outreach success.

Conclusion

Next Steps

As we reach the end of AI & You, consider this not the end, but the beginning of a new chapter in your journey with AI. This book is designed as a living document—one you can return to as your understanding of AI deepens. Equip yourself with a highlighter, sticky notes, and a pen. Personalize your copy by marking passages that resonate, jotting down thoughts, and recording questions that arise.

Insight to Action: Bringing AI & You into Your Daily Life

The real power of AI lies not just in understanding its potential, but in thoughtfully applying it to your life. Turning knowledge into action requires clear goals, practical experimentation, and continual refinement. Here's a simple framework to help bring the lessons of AI & You to life:

1. Set Specific, Measurable, and Time-Bound Goals

Define what you want to achieve with AI and make it measurable. For example, as a manager, you might set a goal like: "Within the next three months, I'll implement AI tools to automate three routine tasks, saving my team five hours of labor per week." A clear, achievable target helps focus your efforts and measure AI's real impact on your workflow.

2. Apply AI in Real-Life Situations

Begin by identifying repetitive tasks that consume time, such as report generation, financial reviews, or meeting transcriptions. Consult the "Case Studies for Career" chapter for ideas on implementing AI one task at a time. Track time savings and gather feedback on each tool's usability and effectiveness.

3. Reflect and Refine

After a few weeks, evaluate the impact of each AI initiative. Determine whether it meets your goals and adjust as needed—whether by fine-tuning the tool, exploring a different solution, or providing further training. As AI becomes part of your team's culture, continue looking for new ways it can support your goals.

By following these steps, you'll integrate AI & You's lessons in a practical way, building both productivity and a workplace that leverages AI for meaningful results.

Leading with AI: Using AI & You with Your Team

Leaders—whether in business, family, or faith—have a unique opportunity to guide others in using AI thoughtfully and responsibly. Beyond enhancing productivity, AI can become a tool that strengthens relationships and purpose.

As a business leader, using AI & You as a team workbook creates a shared understanding of AI's benefits. Equipping team members with their own copies facilitates discovery of AI applications suited to their roles, encouraging responsible AI adoption across your organization. Imagine your team freeing themselves from monotonous tasks, gaining time to focus on creative strategy, organizational agility, and the future.

Conclusion: Next Steps

For family leaders, AI & You can offer guidance on responsible AI use to reinforce family values rather than erode them. Used intentionally, AI can be a positive force for connection and learning. Explore AI with your family, learning together while upholding the values that keep you connected.

For faith leaders, AI & You serves as a resource to explore how AI can enrich personal and community spiritual practices. Whether through deepening prayer, enhancing scripture study, or reaching out to those in need, AI can support what matters most in faith without compromising what is sacred.

Preparing for the Future with AI & You

The pace of AI's development means that while some predictions will materialize, others may take unexpected turns. Rather than seeing these as definitive, use the following projections as prompts to inspire reflection and preparation for what may come:

1. The AGI "Lull"

The push toward Artificial General Intelligence (AGI) will bring excitement and skepticism. As with past technologies, a period of disillusionment often precedes breakthroughs. If true AGI isn't achieved by 2026, expect a temporary pause in momentum as exaggerated claims are met with skepticism. Stocks will fall. Rather than being discouraged, we encourage you to see this lull as an opportunity to refine your AI skills and strategies, positioning yourself for the eventual resurgence.

2. Job Market Disruption

As AI continues automating routine tasks, job disruption is inevitable High-performers across industries are already using AI to outperform their peers. Embrace hands-on experience with AI to future-proof your career, focusing on skills that enable you to solve real-world problems with AI. Developing expertise in AI tools will give you a competitive edge, solidifying your role in a rapidly evolving job market.

3. Growing Energy Demands

AI's increasing energy demands will influence policy changes and public discourse. As society grapples with sustainable energy sources—whether nuclear, renewable, or traditional—prepare by considering diversified energy options for your home and business. Growing resilience in your energy sources can be a valuable step in a world where power reliability is increasingly tied to AI.

4. Big Tech's Role in AI

A handful of tech giants—NVIDIA, Microsoft/OpenAI, Amazon, Google, Meta, and Apple—are leading the race toward AGI. Unlike previous technological developments, AGI's complexity and cost make it accessible only to a few entities. Pay attention to these companies' initiatives, as their choices will reveal both the promise and risks of AI's future.

5. The Race for Artificial General Intelligence (AGI)

The pursuit of AGI—the next frontier in AI—has become a modern-day technological gold rush. Unlike previous innovations, AGI development demands immense resources, including billions of dollars and elite infrastructure, accessible to only a handful of tech giants and superpowers. Companies like NVIDIA, Microsoft/OpenAI, Amazon, Google, Meta, and Apple are leading this race, with initiatives that could reshape society's relationship with technology and power.

Conclusion: Next Steps

6. Rise of MAGA Populism and Trump Mandate

As this book goes to print, Donald J. Trump has won the 2024 Election, gaining a mandate to reshape federal accountability to taxpayers. Initiatives like the Department of Government Efficiency (DOGE) with Elon Musk and Vivek Ramaswamy, alongside a Make America Healthy Again partnership with Robert Kennedy Jr., signal Trump's intent to confront the unelected Administrative State, the Military Industrial Complex, and the excesses of big business on regulation, spending, and foreign policy. This could lead to volatility—both real and imagined—affecting the future of AI.[87, 88, 89]

Looking Forward

In an era saturated with headlines and hype, it's tempting to rely on simplified narratives about AI. Resist the urge. Engage with a variety of perspectives to build a balanced understanding of AI's progress and impact.

By remaining open-minded and informed, you can navigate the evolving AI landscape thoughtfully. Stay vigilant for freedom of speech and respect differing viewpoints—your discernment in identifying reliable information will be invaluable as AI becomes more integrated into our lives.

Finally, remember that personal growth is as important as technological advancement. The future may be complex, but it is also rich with potential for those who engage thoughtfully. As you continue your journey, let curiosity and courage guide you in discovering how AI can enhance your career, family, and spiritual life.

Coming Soon from Zander Curtis

Non-Fiction

Omnipotnent AI: The Moment AI Meets God

Fiction

Blue-Eyed Jesus: AI, Armageddon, & The Rise of the Gød Machine

Endnotes

[1] Stephen Hawking auote found on YouTube: Cambridge University Channel, "The Best or Worst Thing to Happen to Humanity" (2016, October 19), YouTube video, 5:24, https://www.youtube.com/watch?v=_5XvDCjrdXs&t=2s.

Introduction

[2] Reuters, "ChatGPT becomes the fastest-growing consumer app in history," Reuters, February 1, 2023, https://www.reuters.com/technology/chatgpt-becomes-fastest-growing-consumer-app-history-report-2023-02-01/

[3] Center for AI Safety. "Statement on AI Risk." Center for AI Safety. Accessed September 11, 2024. https://www.safe.ai/work/statement-on-ai-risk.

[4] Future of Life Institute. "Pause Giant AI Experiments: An Open Letter." Future of Life Institute. Accessed September 11, 2024. https://futureoJife.org/open-letter/pause-giant-ai-experiments/

[5] White House. "Fact Sheet: President Biden Issues Executive Order on Safe, Secure, and Trustworthy Artifcial Intelligence." The White House, October 30, 2023. https://www.whitehouse.gov/brie7ng-room/statements-releases/2023/10/30/fact-sheet-president-biden-issues-executive-order-on-safe-secure-and-trustworthy-arti7cial-intelligence/.

[6] Booth, Harry, and Tharin Pillay. "Timeline of Recent Accusations Leveled at OpenAI, Sam Altman." Time Magazine, June 7, 2024.https://time.com/6986711/openai-sam-altman-accusations-controversies-timeline/

[7] Ibid

[8] Naomi Buchanan, "As Big Tech Ramps Up AI Spending, Investors Worry Whether AI Costs Will Pay O%," Investopedia, July 3, 2024, https://www.investopedia.com/as-big-tech-ramps-up-ai-spending-investors-worry-whether-costs-will-pay-o%-~669532z:q:text=Even'20once'20realiMed'2C'20the'20outcome,E'uity'20Research'20Head'208im'20Covello

[9] "Breaking Down the Tech Giants— AI Spending Surge," MSN, accessed November 1, 2024, https://www.msn.com/tech-giants-ai-spending.

[10] The Australian Strategic Policy Institute (ASPI). ASPI's Two-Decade Critical Technology Tracker. Canberra: ASPI, September 11, 2023. Accessed Sept 11, 2024. https://www.aspi.org.au/report/aspis-two-decade-critical-technology-tracker.

CONCEPTS

[11] Zander Curtis. Blue-Eyed Jesus: AI, Armageddon, & the Rise of the God Machine.

[12] "AI and Computing Power: The Exponential Growth and Its Impact on Technology." MIT Initiative on the Digital Economy. Last modified January 2023. https://ide.mit.edu/research/importance-exponentially-more-computing-power

1 - What is Artificial Intelligence?

[13] "Microsoft, Meta, Google, and Nvidia Battle for AI Dominance," The Verge, accessed November 1, 2024, https://www.theverge.com/nvidia-ai-hardware

[14] "Artificial Superintelligence and the Singularity." Wikipedia. Last modified February 2024. https://en.wikipedia.org/wiki/Technological_singularity

[15] Goertzel, Ben. "Artificial General Intelligence (AGI) May Arise in 2027 with Artificial 'SuperIntelligence' Sooner Than We Think." Live Science. Last modified March 2023. https://www.livescience.com/ai-singularity-predictions

2 - What is a Human?

[16] Frankl, Viktor E. Man's Search for Meaning. Boston: Beacon Press, 2006.

[17] Robotics Krauss, Patrick. Artificial Intelligence and Brain Research: Neural Networks, Deep Learning and the Future of Cognition. Berlin: Springer, 2024.

[18] Plug and Play. "IoT Sensors Advancing the 5 Human Senses." Plug and Play Tech Center, accessed November 8, 2024. https://www.plugandplaytechcenter.com/insights/iot-sensors-advancing-the-5-human-senses.

[19] Hurt, Avery. "AI and the Human Brain: How Similar Are They?" Discover Magazine, January 14, 2023.

[20] Pascal, Blaise. Pensées. Translated by A. J. Krailsheimer. London: Penguin Classics, 1995.

[21] Plumb. God-Shaped Hole. Curb Records, 1999. CD.

[22] Vogler, Christopher. The Writer's Journey: Mythic Structure for Writers. 3rd ed. Studio City, CA: Michael Wiese Productions, 2007.

[23] Warren, Rick. The Purpose Driven Life: What on Earth Am I Here For? Grand Rapids, MI: Zondervan, 2002.

[24] Campbell, Joseph. The Hero with a Thousand Faces. 3rd ed. Novato, CA: New World Library, 2008.

[25] Douglas Adams, The Hitchhiker's Guide to the Galaxy (London: Pan Books, 1979).

3 - What is Transhumanism?

[26] Curtis, Zander. Red Dragon: AI, Silk Road 3, & The Chinese Dream. Texas, 2026.

[27] Ibid.

[28] Dante Alighieri. The Divine Comedy. Translated by Allen Mandelbaum. New York: Bantam Books, 1982.

[29] Nietzsche, Friedrich. Thus Spoke Zarathustra: A Book for All and None. Translated by Walter Kaufmann. New York: Zodern Library, 1995

Endnotes

30. Huxley, Aldous. Brave New World. New York: Harper & Brothers, 1932

31. Allen, Joe. Dark Aeon: Transhumanism and the War Against Humanity (New York: War Room Books, 2023)

32. Ibid.

33. Harari, Yuval Noah. Homo Deus: A Brief History of Tomorrow. New York: Harper Collins, 201#.

34. Ibid.

35. Regalado, Antonio. "The First Gene-Editing Treatment: 10 Breakthrough Technologies 2024." MIT Technology Review, January 8, 2024. https://www.technologyreview.com/2024/01/08/1085101/crispr-gene-editing-sickle-cell-disease-breakthrough-technologies/.

36. Scharre, Paul. Four Battlegrounds: Power in the Age of Artificial Intelligence. New York: W.W. Norton & Company, 2023.

37. Morgan, Forrest E., et al. "Military Applications of Artificial Intelligence: Ethical Concerns in an Uncertain World." RAND Corporation, April 28, 2020.

38. Center for AI Safety. "Statement on AI Risk." Center for AI Safety. Accessed September 11, 2024. https://www.safe.ai/work/statement-on-ai-risk.

39. Strickland, Eliza. "IBM Watson, Heal Thyself: How IBM Overpromised and Underdelivered on AI Health Care." IEEE Spectrum, April 2019. (IEEE Spectrum)

40. United States. Scientific and Advanced Technology Act of 1992. Public Law 102-476. U.S. Statutes at Large 106 (1992): 2297–2300.

4 - The First Convergence

41. OpenAI. ChatGPT-4. Definition of "Convergence." Accessed November 8, 2024. https://chat.openai.com/.

42. Apple Inc. "Steve Jobs Introduces the iPhone." Keynote address, Macworld Conference & Expo, San Francisco, CA, January 9, 2007. https://www.apple.com/steve-jobs/iphone-launch/.

43. Vaswani, Ashish, Noam Shazeer, Niki Parmar, Jakob Uszkoreit, Llion Jones, Aidan N. Gomez, Lukasz Kaiser, and Illia Polosukhin. "Attention Is All You Need." Advances in Neural Information Processing Systems 30 (2017): 5998–6008. https://papers.nips.cc/paper/7181-attention-is-all-you-need.pdf.

44. United States. Scientific and Advanced Technology Act of 1992. Public Law 102-476. U.S. Statutes at Large 106 (1992): 2297–2300.

45. Weinberger, Sharon. The Imagineers of War: The Untold Story of DARPA, the Pentagon Agency That Changed the World. New York: Knopf, 2017.

[46] Mazzucato, Mariana. "The Entrepreneurial State." Soundings 49 (2011): 131–142.

[47] Defense Advanced Research Projects Agency. "DARPA Joins Public-Private Partnership to Address Challenges Facing Microelectronics." DARPA News, January 21, 2021.

5 - The Second Convergence

[48] Fridman, L., & Altman, S. (2024). Lex Fridman Podcast #419: Sam Altman [Video]. YouTube. https://www.youtube.com/watch?v=L_Guz73e6fw

[49] Ibid.

[50] Ibid.

[51] Center for AI Safety. "Statement on AI Risk." Center for AI Safety. Accessed September 11, 2024. https://www.safe.ai/work/statement-on-ai-risk.

[52] Future of Life Institute. "Pause Giant AI Experiments: An Open Letter." Future of Life Institute. Accessed September 11, 2024. https://futureoJife.org/open-letter/pause-giant-ai-experiments/

[53] Elon Musk, meeting with U.S. Senators, September 14, 2023. The Independent. Accessed August 2, 2024. The Independent .

[54] Fridman, L., & Altman, S. (2024). Lex Fridman Podcast #419: Sam Altman [Video]. YouTube. https://www.youtube.com/watch?v=L_Guz73e6fw

[55] (99) Russell, Stuart. Human Compatible: Artificial Intelligence and the Problem of Control. New York: Viking, 2019.

[56] Ibid.

[57] Mark Zuckerberg quoted in Andrew Griffin, "Mark Zuckerberg wants Facebook's AI systems to help doctors spot diseases," The Independent, February 16, 2017.

[58] Zuckerberg Mark."Integrating Artificial Intelligence: Future Implications." Journal Social Media Research Volume IX Issue V November (2022)

6 - The Third Convergence

[59] Zander Curtis. Alice: AI, Blue Butterfly, & Conspiracy Theory No.4. (Tranquil Pause Publishing, 2026).

[60] Ibid.

[61] Nick Bostrom, Superintelligence: Paths, Dangers, Strategies (Oxford: Oxford University Press, 2014).

[62] Yuval Noah Harari, Homo Deus: A Brief History of Tomorrow (New York: HarperCollins Publishers Inc., 2017)

[63] Ray Kurzweil, The Singularity Is Near: When Humans Transcend Biology (New York: Viking, 2005).

Endnotes

[64] Ibid.

[65] Max Tegmark, Life 3.0: Being Human in the Age of Artificial Intelligence (New York: Knopf Publishing Group, 2020)

[66] Stephen Hawking, interview, The Independent, May 1, 2014, accessed August 2, 2024. https://www.independent.co.uk/news/science/stephen-hawking-ai-biggest-event-human-history-9321547.html .

[67] Stephen Hawking, interview, Wired, November 2, 2017, accessed August 2, 2024. https://www.wired.co.uk/article/stephen-hawking-interview-ai-viruses.

[68] Tegmark, Max. Life 3.0: Being Human in the Age of Artificial Intelligence. New York: Knopf, 2017.

[69] Frey, Carl Benedikt. The Technology Trap: Capital, Labor, and Power in the Age of Automation. Princeton University Press.

[70] Ibid.

[71] Kaku, Michio. The Future of Humanity: Terraforming Mars, Interstellar Travel, Immortality, and Our Destiny Beyond Earth. New York: Doubleday, 2018

7 - YOUR AI COMPANION

[72] McStay, Andrew. Emotional AI: The Rise of Empathic Media. Sage Publications Ltd., 2018.

[73] Borenstein, Jason, and Yvette Pearson. "The Ethics of Artificial Intelligence in Healthcare: Should AI Companions Be Allowed to Make Ethical Decisions?" The AZA Journal of Ethics 23, no. 10 (2021): E792-798

[74] Glover, Ellen. "What Are AI Agents?" Built In, October 31, 2024. https://builtin.com/articles/ai-agents

[75] Amazon Web Services. "What Are AI Agents?" Accessed November 8, 2024. https://aws.amazon.com/what-is/ai-agents/

[76] Wooldridge, Michael. An Introduction to MultiAgent Systems. 2nd ed. Chichester: John Wiley & Sons, 2009

[77] "Speech Recognition Faster at Texting," Stanford University, August 24th, 2016. https://news.stanford.edu/stories/2016/08/stanford-study-speech-recognition-faster-texting

8 - MEETING YOUR AI COMPANION

[78] Russell, Stuart J., and Peter Norvig. Artificial Intelligence: A Modern Approach. 4th ed. Upper Saddle River: Pearson, 2020.

[79] McKinsey & Company. "What Is Prompt Engineering?" McKinsey & Company, March 22, 2024. https://www.mckinsey.com/featured-insights/mckinsey-explainers/what-is-prompt-engineering.

9 - Talking to Your AI Companion

[80] Prompt Engineering Guide. "Prompt Engineering Guide." Accessed November 8, 2024. https://www.promptingguide.ai/.

10 - Protect Yourself From Your AI Companion

[81] WIRED. "In Defense of AI Hallucinations." WIRED, January 2024. https://www.wired.com/story/plaintext-in-defense-of-ai-hallucinations-chatgpt/.

[82] Ibid.

[83] IBM. "What Are AI Hallucinations?" Accessed November 8, 2024. https://www.ibm.com/topics/ai-hallucinations.

[84] Binns, Reuben. "Fairness in Machine Learning: Lessons from Political Philosophy." In Proceedings of the 2018 Conference on Fairness, Accountability, and Transparency, 149-159. New York: Association for Computing Machinery, 2018. https://doi.org/10.1145/3287560.3287584

[85] Campolo, Alex, Madelyn Sanfilippo, Meredith Whittaker, and Kate Crawford. "AI Now Report 2017." New York: AI Now Institute, 2017. https://ainowinstitute.org/AI_Now_2017_Report.pdf.

[86] OpenAI. "Our Approach to AI Safety." OpenAI, April 5, 2023. https://openai.com/index/our-approach-to-ai-safety/

Conclusion: Next Steps

[87] "Trump, Elon Musk, Vivek Ramaswamy Discuss Government Efficiency and the 'Deep State'," NPR, November 12, 2024, https://www.npr.org/2024/11/12/g-s1-33972/trump-elon-musk-vivek-ramaswamy-doge-government-efficiency-deep-state.

[88] "Robert F. Kennedy Jr. Named as Trump's HHS Secretary Pick," Politico, November 14, 2024, https://www.politico.com/news/2024/11/14/robert-f-kennedy-jr-trump-hhs-secretary-pick-00188617.

[89] Louis Casiano, "Dem Rep. Robert Garcia Says RFK Jr. Nomination for Health Secretary Is 'F—— Insane,' Will 'Cost Lives'," Fox News, November 14, 2024, https://www.foxnews.com/politics/dem-rep-robert-garcia-says-rfk-jr-nomination-health-secretary-f-insane-cost-lives.

Resources

Thought Leaders

- Aristotle (philosopher, logician) - Acieent work in logic and metaphysics which provides foundational principles for modern AI reasoning systems.
- Joe Allen (author, transhumanism critic) - Critiques the implications of merging humans with technology and the dangers to human identity.
- Isaac Asimov (author, futurist) - Known for his Three Laws of Robotics, Asimov's science fiction work has influenced how society views AI ethics.
- Nick Bostrom (philosopher, author) - A philosopher who focuses on AI safety, existential risks, and the potential dangers of superintelligence.
- Joseph Campbell (mythologist, writer) - Known for heroes journey concept and mythology offering frameworks for understanding human storytelling.
- Zander Curtis (author, futurist) - Delves into the intersection of AI, society, and spiritual beliefs in his works, blending fiction and non-fiction.
- Pedro Domingos (computer scientist, AI researcher) - A machine learning researcher who authored The Master Algorithm.
- Sigmund Freud (neurologist, psychoanalyst) - His theories on the human mind offer explanations for human consciousness and decision-making.
- Ben Goertzel (AI researcher, computer scientist) - Goertzel's AGI work focuses on designing systems that achieve ethical general intelligence.
- Stephen Hawking (theoretical physicist, cosmologist) - Warned of the existential risks posed by AGI and ASI, emphasizing human control.
- Carl Jung (psychiatrist, psychoanalyst) - Work on the human psyche and collective unconscious is being applied to developing human thinking in AI.
- Daniel Kahneman (psychologist, economist) - Known for his work on decision-making and behavioral economics which AI pioneers closely follow.
- Kai-Fu Lee (AI researcher, entrepreneur) - Explores the geopolitical and economic impacts of AI, especially the competition between China and the U.S.
- Ray Kurzweil (futurist, inventor) - A frequently cited forecaster and advocate of the Singularity, a future point where AI exceeds human intelligence.
- Jaron Lanier (computer scientist, virtual reality pioneer) - A pioneer in virtual reality and a critic of AI's potential social impact.

- » C.S. Lewis (writer, theologian) - His exploration of Christian apologetics and human nature spurn discussions how AI impacts morality and faith.
- » Marvin Minsky (cognitive scientist, AI researcher) - Founding father of AI, his work in cognitive science and robotics shape principles of modern AI.
- » Martha Nussbaum (philosopher, ethicist) - Work provides an ethical lens through which AI development can be critiqued.
- » Stuart Russell (AI researcher, computer scientist) - One of the leading voices in AI safety and ethics, Russell has proposed strategies for ensuring AI systems are aligned with human values and remain under control.
- » Alan Turing (mathematician, logician) - Pioneer thinker for modern computing who developed foundational ideas in AI such as the Turning Test.
- » Max Tegmark (physicist, AI researcher) - Explores the future of intelligence, advocating for AI safety and ethical development of advanced AI systems.
- » Vernor Vinge (author, mathematician) - A mathematician and science fiction writer, Vinge is credited with popularizing the concept of the technological Singularity, where AI surpasses human intelligence.

Books

Recommended Books

- » Homo Deus: A Brief History of Tomorrow (2016) by Yuval Noah Harari explores future scenarios where humans use AI and biotechnology to transcend biological limitations, raising ethical and existential questions.
- » Man's Search for Meaning (1946) by the author Viktor Frankl shares his experiences in Nazi concentration camps and his psychological insights on finding purpose in life, even amid suffering.
- » Life 3.0: Being Human in the Age of Artificial Intelligence (2017) by Max Tegmark examines the future of AI, discussing the ethical, social, and existential questions raised by the development of intelligent machines.
- » AI Superpowers: China, Silicon Valley, and the New World Order (2018) by Kai-Fu Lee compares the AI strategies of China and the U.S., highlighting the geopolitical race to dominate AI innovation.

Resources 255

- » Dark Aeon: Transhumanism and the War Against Humanity (2023) by Joe Allen critiques the growing transhumanism movement including consequences of merging humanity with machines.
- » Antifragile: Things That Gain from Disorder (2012) by Nassim Taleb explains how some things grow stronger from chaos and uncertainty.
- » Social Physics: How Good Ideas Spread (2014) by Alex Pentland explores how social networks and data influence the spread of ideas and how these processes can be harnessed to understand collective behavior.
- » Guns, Germs, and Steel: The Fates of Human Societies (1997) by Jared Diamond examines the factors that have shaped the development of civilizations.
- » Sapiens: A Brief History of Humankind (2014) by Yuval Noah Harari traces the evolution of human societies and considers how AI might shape the future of humanity.
- » Our Final Invention: Artificial Intelligence and the End of the Human Era (2013) by James Barrat investigates the risks associated with AI development, highlighting the dangers of machines that outthink their creators.
- » Blue-Eyed Jesus: AI, Armageddon & The Rise of the God Machine (Coming 2025) by Zander Curtis explores the intersection of AI technology, theology, and the concept of an apocalyptic future driven by machines.

Other Books

- » The AI Apocalypse (2023) by Jonas Salk explores the potential dangers of AI, focusing on the risks that emerging technologies pose to humanity and offering strategies for mitigating these risks.
- » Reality+: Virtual Worlds and the Problems of Philosophy (2022) by David Chalmers examines the intersection of AI, virtual reality, and philosophical questions about the nature of existence and reality.
- » Atlas of AI: Power, Politics, and the Planetary Costs of Artificial Intelligence (2021) by Kate Crawford critiques the ethical and environmental consequences of AI development and the power structures that drive the industry.
- » The Age of AI: And Our Human Future (2021) by Henry A. Kissinger, Eric Schmidt, and Daniel Huttenlocher explores the geopolitical, ethical, and societal impacts of AI, with a focus on the balance of innovation and control.

- » Apocalypse Never: Why Environmental Alarmism Hurts Us All (2020) by Michael Shellenberger challenges common environmental narratives, arguing that technological progress can solve environmental issues without alarm.
- » Aristotle's Logic: An Introductory Guide (2020) by Robin Smith serves as an introduction to Aristotle's work on logic, drawing connections between ancient philosophy and modern computational systems.
- » Machine, Platform, Crowd: Harnessing Our Digital Future (2017) by Erik Brynjolfsson and Andrew McAfee delves into how digital technologies are transforming business models and social structures.
- » Weapons of Math Destruction: How Big Data Increases Inequality and Threatens Democracy (2016) by Cathy O'Neil investigates the dark side of big data and how algorithms can exacerbate societal inequality.
- » The Master Algorithm: How the Quest for the Ultimate Learning Machine Will Remake Our World (2015) by Pedro Domingos outlines the quest to create a universal learning algorithm capable of driving the next wave of AI.
- » The Oxford Handbook of Apocalyptic Literature (2014) edited by John J. Collins offers an academic exploration of apocalypticism throughout history, examining its influence on culture, religion, and society.
- » The Fourth Revolution: How the Infosphere is Reshaping Human Reality (2014) by Luciano Floridi examines the philosophical implications of living in the 'infosphere,' a digital reality shaped by AI and information technology.
- » The Printing Revolution in Early Modern Europe (1983) by Elizabeth L. Eisenstein explores how the printing press revolutionized communication, with parallels to how AI is transforming information dissemination today.
- » Black Death: A Personal History (2008) by John Hatcher provides a detailed account of the Black Death in England, offering historical context for discussions about pandemics, technology, and societal change.
- » The Black Swan: The Impact of the Highly Improbable (2007) by Nassim Nicholas Taleb explores the role of rare, unpredictable events in shaping history and technology, with implications for AI safety.
- » The Great Mortality: An Intimate History of the Black Death, the Most Devastating Plague of All Time (2005) by John Kelly chronicles the Black Death, drawing parallels to modern global challenges.

- » The Age of Spiritual Machines: When Computers Exceed Human Intelligence (1999) by Ray Kurzweil predicts the rise of AI and its potential to surpass human intelligence, exploring philosophical and societal implications.
- » Gutenberg: How One Man Remade the World with Words (2002) by John Man traces the history of the printing press and its parallels to modern technological revolutions like AI.
- » Machines of Loving Grace: The Quest for Common Ground Between Humans and Robots (2015) by John Markoff explores the relationship between humans and machines, focusing on the ethical dilemmas posed by AI.
- » The Digital Deluge (2022) by Lila Moore discusses the effects of digital technology on society and how AI is reshaping humanity and spirituality.
- » End Times and Tech Giants (2022) by Eric Newstadt explores the convergence of AI development with apocalyptic themes.
- » When Robots Pray (2021) by Fiona Murphy examines the intersection of AI and spirituality, exploring how AI technologies might challenge or complement religious practices.
- » Brand Luther: 1517 Printing and the Making of the Reformation (2015) by Andrew Pettegree examines the role of the printing press in shaping the Reformation, drawing parallels to the technological revolutions of today.
- » Human Compatible: Artificial Intelligence and the Problem of Control (2019) by Stuart Russell addresses the ethical and technical challenges of building AI systems that are aligned with human values and goals.
- » Leviathan and the Air-Pump: Hobbes, Boyle, and the Experimental Life (1985) by Steven Shapin and Simon Schaffer recounts the historical debates about scientific experimentation, drawing analogies to modern AI.
- » The Singularity Is Near: When Humans Transcend Biology (2005) by Vernor Vinge predicts the convergence of biological and artificial intelligence, offering a vision of the future where humanity may transcend its biology.
- » The Selfish Gene (1976) by Richard Dawkins presents a gene-centered view of evolution, exploring how biological mechanisms may have parallels in AI systems designed to 'evolve' and optimize.

PODCASTS

- The Joe Rogan Experience – Top podcast in the world not afraid of exploring forbidden topics with guests from culture, politics, sports, AI, and science.
- The Lex Fridman Podcast – As an AI researcher, Lex dives deep into conversations with prominent thinkers focusing on AI, philosophy, and humanity.
- The David Shapiro AI Podcast – Explores the latest advancements in artificial intelligence and the societal implications.
- The All In Podcast, hosted by tech entrepreneurs and investors Chamath Palihapitiya, Jason Calacanis, David Sacks, and David Friedberg, offers insightful discussions on business, technology, politics, and current events.
- Hidden Brain - Explores the unconscious patterns driving human behavior, combining storytelling, science and how we interact with each other.
- Impact Theory with Tom Bilyeu – Inspirational interviews with thought leaders, focusing on personal development, mindset, and strategies
- Modern Wisdom with Chris Williamson – Featuring interview he refers to as "life lessons from the smartest people on the planet."
- The Tim Ferriss Show – Tim Ferriss invites experts from various fields to share insights, routines, and strategies for enhancing productivity, mental resilience, and personal growth.
- Huberman Lab Podcast – Neuroscientist Dr. Andrew Huberman breaks down complex neuroscience topics.
- The Artificial Intelligence Show with Paul Roetzer and Mike Kaput - Explores how AI is transforming business, providing strategies for professionals to leverage AI-driven tools.
- The AI Podcast by NVIDIA - Insights from experts on AI's advancements and applications across industries.
- The AI Brief – Concise updates on the latest AI developments, emerging applications and trends.
- Everyday AI – Explores practical AI applications in daily life and work, featuring interviews and real-world examples.
- The Future of Humanity Podcast – Hosted by Nikola Danaylov, this podcast delves into transhumanist topics, including human enhancement.

Resources

- » Hacking Humans – While it primarily focuses on cybersecurity and human behavior, this podcast frequently touches on transhumanist themes and the ethical considerations of enhancing human abilities through technology.
- » The Singularity Podcast – Focuses on futurist technological advancements, covering topics such as artificial intelligence, brain-computer interfaces, and human-machine integration.
- » Brain Inspired Podcast – Explores neuroscience and AI, discussing how neural technologies and brain-machine interfaces are paving the way for deeper human-machine integration.
- » Future of Life Institute Podcast – Covers a broad spectrum of future technologies, including AI, and their potential risks to humanity, with a focus on existential risk and ethical considerations.
- » Data Skeptic – Delves into the technical aspects of AI, machine learning, and data science, while addressing the societal challenges these technologies pose.

MOVIES

Recommended Movies

- » Her (2013) – Follows a man who develops a deep emotional relationship with an AI operating system, raising questions about AI and consciousness.
- » Minority Report (2002) – Set in a world where predictive AI technology is used to prevent crimes before they happen, this movie examines the moral implications of preemptive justice and surveillance.
- » The Matrix (1999) – Depicts a future where humanity is enslaved by AI machines that control reality through a simulated world, questioning the nature of reality and human autonomy.
- » The Terminator (1984) – Features an apocalyptic scenario where an AI defense system, Skynet, turns against humanity, highlighting the dangers of autonomous AI systems.
- » Bicentennial Man (1999) – Chronicles the journey of an AI robot over two centuries as it strives to become more human, highlighting themes of AI identity and human-like evolution.
- » 2001: A Space Odyssey (1968) – Depicts HAL 9000, an AI system that demonstrates advanced capabilities and potential dangers of AI autonomy.

Other AI-Related Movies

- » The Creator (2023) – Set in a dystopian future where AI has reached human-level intelligence, the film explores the moral complexities of war between humans and AI-driven robots.
- » M3GAN (2023) – A horror-thriller about a life-like AI doll designed to protect a child, raising concerns about AI autonomy and the ethical risks of creating AI companions for children.
- » Finch (2021) – Starring Tom Hanks, this film tells the story of a man, his dog, and a humanoid robot as they navigate a post-apocalyptic world, exploring AI's role in human survival and companionship.
- » Outside the Wire (2021) – A military sci-fi thriller where a drone pilot teams up with an advanced AI-driven android soldier, exploring AI's role in warfare and the ethical dilemmas of autonomous military systems.
- » Ava (2020) – A thriller about an AI assassin program that begins to question its purpose, addressing the concerns of autonomous AI weapon systems.
- » The Social Dilemma (2020) – Discusses the use of AI algorithms by social media platforms to influence human behavior and decision-making, raising concerns about privacy and manipulation.
- » Transcendence (2014) – Follows a scientist who uploads his consciousness into an AI system, sparking debates about human identity, immortality, and AI's potential power.
- » Ex Machina (2014) – Explores the ethical implications of creating highly advanced AI with human-like consciousness and emotional intelligence.
- » The Imitation Game (2014) – Focuses on the life of Alan Turing and his work on breaking the Enigma code, laying the foundation for modern computing and AI.
- » Eagle Eye (2008) – In this thriller, an AI system controls a vast network of data and technology, manipulating humans to prevent perceived threats, showcasing the potential risks of AI surveillance and control.
- » I, Robot (2004) – Examines the potential consequences of AI robots developing free will and the conflicts that arise between safety and AI autonomy.
- » A.I. Artificial Intelligence (2001) – Explores the emotional and ethical impli-

cations of creating AI with human-like qualities.
- » Ghost in the Shell (1995) – Focuses on the blurred line between humans and AI in a the future where consciousness can be transferred to machines.
- » Tron (1982) – A classic film about a computer programmer who is transported into a digital world where AI programs fight for dominance, reflecting on the evolving relationship between humans and technology.
- » Blade Runner (1982) – Set in a dystopian future where AI "replicants" are nearly indistinguishable from humans, raising questions about identity, consciousness, and AI ethics.
- » Westworld (1973) – Set in a futuristic theme park where AI robots malfunction, exploring the concerns surrounding AI, consciousness, and control.

TV Shows

AI-Related Shows

- » "The Orville" (2017-present) – Explores AI ethics and consequences of advanced AI, featuring the AI species known as the Kaylon.
- » "Star Trek: Discovery" (2017-present) – Portrays advanced AI systems like "Control," posing existential threats to organic life.
- » "Black Mirror" (2011-present) – Examines AI, technology, and its impacts on society through dark, dystopian narratives.
- » "NeXt" (2020) – Follows the rise of a rogue AI system and the struggle to prevent its destructive potential.
- » "Person of Interest" (2011-2016) – Centers on an AI system for mass surveillance, exploring ethics of predictive policing.

Being Human Related Shows

- » "The Handmaid's Tale" (2017-present) – Examines societal control and dystopian themes, with technology as a tool of oppression.
- » "Medici: Masters of Florence" (2016-2019) – Depicts historical impacts of innovation during the Renaissance.
- » "Mr. Robot" (2015-2019) – Focuses on hacking, surveillance, and societal structures, reflecting on human vulnerability in a tech-dominated world.

- » "Revolution" (2012-2014) – Shows humanity's struggle to rebuild society in a world without technology.
- » "The X-Files" (1993-2018) – Blends science, technology, and paranormal investigations, focusing on human curiosity and fear of the unknown.

Transhuman/Android-Related Shows

- » "Star Trek: Picard" (2020-present) – Delves into synthetic life, AI rights, and the future of sentient AI beings.
- » "Raised by Wolves" (2020-2022) – Explores androids raising human children and themes of human evolution.
- » "Westworld" (2016-2022) – Explores AI, consciousness, and society's decay within a theme park populated by android hosts.
- » "Humans" (2015-2018) – Examines AI's societal challenges as human-like robots, or "synths," integrate into everyday life.
- » "Battlestar Galactica" (2004-2009) – Focuses on human vs. AI Cylons, exploring themes of survival, identity, and humanity.
- » "Altered Carbon" (2018-2020) – Explores transferring human consciousness between bodies, blending themes of AI, technology, and human identity.

Songs

AI/Technology Related Songs

- » "Artificial" by Daughtry (2024)
- » "Algorithm" by Muse (2018)
- » "When The World Was At War We Kept Dancing" by Lana Del Rey (2017)
- » "Spit Out the Bone" by Metallica (2016)
- » "A.I." by OneRepublic featuring Peter Gabriel (2016)
- » "Radioactive" by Imagine Dragons (2012)
- » "Robots" by Flight of the Conchords (2005)
- » "Technologic" by Daft Punk (2005)
- » "Paranoid Android" by Radiohead (1997)
- » "Mr. Roboto" by Styx (1983)

- » "Computer Age" by Neil Young (1982)
- » "Video Killed the Radio Star" by The Buggles (1979)

Being Human Releated Songs
- » "Afterlife" by Arcade Fire (2013)
- » "Blackbird" by Alter Bridge (2007)
- » "Bring Me To Life" by Evanescence (2003)
- » "Higher" by Creed (1999)
- » "Losing My Religion" by R.E.M. (1991)
- » "Man in the Mirror" by Michael Jackson (1987)
- » "Hallelujah" by Leonard Cohen (1984)
- » "Bargain" by The Who (1971)
- » "Imagine" by John Lennon (1971)
- » "What's Going On" by Marvin Gaye (1971)
- » "Let it Be" by The Beatles (1970)
- » "Blowin' in the Wind" by Bob Dylan (1962)

INDEX

2001: A Space Odyssey (movie): 9, 27, 257

2084: Artificial Intelligence and the Future of Humanity (book): 27

A.I. Artificial Intelligence (movie): 68, 258

"A.I." by OneRepublic featuring Peter Gabriel (song): 28, 260

AGI (Artificial General Intelligence): 34-39, 71, 83, 88, 91, 94-110, 121, 241-243

Algorithm: 12, 21, 35-41, 59, 66, 69, 77-81, 86, 89, 123, 124, 134, 138, 163, 251, 254, 258, 260

"Algorithms" by Muse (song): 28, 69, 260

Altman, Sam: 3, 26, 93-100

Amazon (company): 80, 85, 88, 134, 215, 242

Apple (company): 62, 81-84, 133, 212

"Artificial" by Daughtry (song): 24, 260

Artificial General Intelligence (AGI): 34-39, 71, 83, 88, 91, 93-110, 121, 241-243, 251

Artificial Narrow Intelligence (ANI): 14, 15, 73, 76, 77 81, 96

Artificial Superintelligence (ASI): 14, 16, 106, 111-114, 116, 121, 123, 125, 129

"Bargain" by The Who (song): 49, 261

Biden, Joe: 3

Bilyeu, Tom: 48, 256

Blade Runner (movie): 48, 68, 259

"Blowin' in the Wind" by Bob Dylan (song): 49, 261

Blüe Butterfly (fictional company): 111-114

Blue Eyed Jesus (book) 7, 253

Bostrom, Nick: 26, 115-116, 251

Brain-Computer Interface (BCI): 55, 59, 63, 64, 68, 98, 120, 257,

Brandt, Zoe: 111-114

Campbell, Joseph: 42, 47, 251

ChatGPT: 1, 3, 15, 71, 82-84, 131-136, 1411-145, 148, 172, 175, 177, 179, 181, 184, 186, 192, 196, 198, 200-205, 207, 209-212, 215, 221, 223, 225, 229, 231, 235

"Closer" by Nine Inch Nails (song): 69

CRISPR (Clustered Regularly Interspaced Short Palindromic Repeats): 60-62

Cybersecurity: 68, 89, 102, 107, 161, 167, 168, 257,

DARPA (government agency): 73-76, 86, 87

Data Privacy: 22, 23, 62-64, 76, 79, 85-89, 90, 106, 108, 126-128, 161, 166, 258

Dawkins, Richard: 48, 255

de Grey, Aubrey: 67

Deep Learning: 19, 34, 81, 85,

DeepMind (company): 61

Diamond, Jared: 47, 255

Ex Machina (movie): 68, 88, 158

Existential Risk: 1-3, 17, 16, 26, 28, 35, 40, 41, 48, 84, 118, 119, 251, 257,

Facebook (company): 80, 143,

Facial Recognition: 82, 138, 163

Frankl, Viktor Emil: 29-32, 48, 252

Freedom of Speech: 3, 37, 76, 107, 122, 164, 243

Fridman, Lex: 27, 93-96, 99, 256

Future of Life Institute: 2, 98, 110, 247, 257

Generative Pre-trained Transformers (GPT): 10, 78, 82-84

Gaye, Marvin: 49, 261

Google (company): 4, 15, 79, 81-85, 88, 117, 133-136, 142, 145, 148, 154, 205, 215

Guns, Germs, and Steel: 47, 253

Harari, Yuval Noah: 2, 47, 58, 84, 116, 252

Hawking, Stephen: 1, 58

"Higher" by Creed (song): 49, 261

Hitchhiker's Guide To The Galaxy (book): 42, 48

Homo Deus (book): 58, 67, 116, 252

Huberman, Andrew: 49, 256

Human Algorithm: 41

Huxley, Aldous: 57

IBM (company): 10, 61, 87, 133

"Imagine" by John Lennon (song): 13

Internet of Things (IoT): 10, 88, 139

Jackson, Michael: 49, 261

Jung, Carl: 47, 251

Kurzweil, Ray: 26, 67, 117, 251, 255

Large Language Models (LLMs): 13, 14, 41, 46, 81, 82, 134-136, 139, 141-145, 147-148, 171-216

Lee, Kai-Fu: 26, 251, 252

Lennon, John: 13

Lennox, John C.: 26, 27

Lewis, C.S.: 47, 48, 252

"Losing My Religion" by R.E.M. (song): 49, 261

Machine Learning (ML): 13, 21, 34, 35, 59, 79, 80, 81, 87, 151, 249, 251, 257

MAGA (Make America Great Again) 243

Meta (company): 4, 59, 100, 142, 143, 148,

Microsoft (company): 3, 15, 85, 134, 136, 142, 143, 148, 177, 181, 187, 196, 200-203

Musk, Elon: 2, 26, 59, 84, 98, 100, 143, 243, 248

Natural Language Processing (NLP): 18, 81, 85, 87, 128, 134

Neural Networks: 10, 13, 33, 81, 151,

Neuralink (company) 59, 98, 120

Nietzsche, Friedrich: 56, 57

NVIDIA (company): 27, 84, 256

Omnipotent AI: 40, 129,

OpenAI (company): 1, 3, 15, 26, 83, 84, 93, 94, 99, 137, 142,

"Paranoid Android" by Radiohead (song): 28, 69, 261

Predictive Analytics: 21, 28, 81, 84, 119, 121, 125, 136, 163, 257, 259

Quantum Computing: 88

Robotics: 86, 87, 103, 111, 112, 119, 126, 135, 137, 138, 252, 255, 257-260

Rogan, Joe: 27, 256

Russell, Stuart: 26, 99, 252, 255

Shapiro, David: 27, 256

Singularity: 26, 67, 86, 117, 252, 255, 257

Siri: 15, 77, 81, 84, 131-137

Index

Synthetic Biology: 9, 33, 61, 62, 68, 260

Tegmark, Max: 26, 117, 118, 252

Tesla (company): 26, 98,

The Matrix (movie): 27, 88, 257 Tesla (company): 26, 98,

The Terminator (movie): 9, 68

Trump, Donald: 243

Transhumanism: 7, 51-68, 251, 253,

Virtual Assistants: 25, 133, 136,137, 215, 216

Virtual Reality (VR): 128, 137, 251, 253

Voice Recognition: 10, 37, 52, 81,

"What's Going On" by Marvin Gaye (song): 49, 251

www.ingramcontent.com/pod-product-compliance
Lightning Source LLC
Chambersburg PA
CBHW040237110526
44582CB00023B/221/J